BRAVE NEW UNIVERSE

ILLUMINATING

THE

DARKEST SECRETS

OF THE

COSMOS

PAUL HALPERN AND PAUL WESSON

Joseph Henry Press
Washington, D.C.

Joseph Henry Press • 500 Fifth Street, NW • Washington, DC 20001

The Joseph Henry Press, an imprint of the National Academies Press, was created with the goal of making books on science, technology, and health more widely available to professionals and the public. Joseph Henry was one of the founders of the National Academy of Sciences and a leader in early American science.

Any opinions, findings, conclusions, or recommendations expressed in this volume are those of the author and do not necessarily reflect the views of the National Academy of Sciences or its affiliated institutions.

Library of Congress Cataloging-in-Publication Data

Halpern, Paul, 1961-
   Brave new universe : illuminating the darkest secrets of the cosmos / by Paul Halpern and Paul Wesson.— 1st ed.
      p. cm.
   Includes bibliographical references and index.
   ISBN 0-309-10137-9 (hardcover : alk. paper) — ISBN 0-309-65823-3 (PDFs : alk. paper) 1. Cosmology. I. Wesson, Paul S. II. Title.
   QB981.H248 2006
   523.1—dc22

                                      2006004464

Cover Image © Mark Garlick/Photo Researchers, Inc.

Printed in the United States of America.

O, wonder! How many goodly atoms are there here! How mysterious the dark substances and energies are! O brave new universe that hath such diverse materials in it!

The Cosmic Tempest
(with apologies to William Shakespeare)

Contents

Preface

In this book the two Pauls wish to share with readers the fascination that is modern cosmology—the study of the universe. Recent years have seen monumental progress in this field, transforming it from a compendium of rough observations and general trends into a finely honed science, analyzed through statistical and computational techniques. This revolution is akin to the leap taken in weather forecasting when meteorologists turned to detailed mathematical models and to the strides in genetics when biologists began mapping out the precise structures of chromosomes.

As with other human pursuits, we can learn a lot about cosmology by looking at its puzzles and seeing how they can be solved. Archaeologists do this when they study the structure of the ancient pyramids. Musicians do it when they pore over the scores of bygone composers. Painters do it when they scrutinize the techniques of the Old Masters. Even physicians do it, when in setting out to cure an illness they develop a better understanding of what it means to be healthy. Our aim is to examine the conundrums posed by cosmology so that, through their resolution, we can obtain a deeper comprehension of the universe.

A typical example is Olbers' paradox—the mystery of why the night sky is dark. Under what appear to be fairly simple assumptions, the universe should be ablaze with the light from trillions of stars and galaxies, instead of the speckled black we observe. Anyone with a clear mind, anyone who has been awed by the darkness of a clear night, can solve this paradox—and thus reveal deep truths about the nature of space. Delving into other basic issues offers valuable insight about additional aspects of reality: Where are the aliens? What would happen if we fell into a black hole? Was there really a Big Bang? Can matter have negative mass? Are there extra dimensions and perhaps parallel universes? Are we as humans affected by the most remote objects we see through large telescopes, such as quasars?

In this book we demonstrate how even in this age of fantastic new technologies—with reams of telescopic data flooding astronomers' computers—some of the most poignant dilemmas can be tackled through sheer reasoning. Often, one does not need fancy machines or complicated mathematics to get to the crux of a subject. An old English saying warns of the danger of not being able to see the wood for the trees. As we will show, our insights help us reveal the "solid wood" of modern cosmology. With all our grand explorations, nothing is more extraordinary than the power of the human mind.

Introduction:
The Quest for Cosmic Understanding

*I am greatly relieved that the universe is finally explainable.
I was beginning to think it was me. As it turns out, physics, like
a grating relative, has all the answers.*

Woody Allen (*The New Yorker*, July 28, 2003)

A short time ago cosmology seemed settled in the comfy chair of complacency, confident in the apparent resolution of many of its major issues. All known data converged on a uniform chronology of cosmic history—an account so widely accepted that it had come to be known as the standard model.

Every student of astronomy could recite the then-known facts: The universe began in a fiery explosion called the Big Bang and then expanded for billions of years. Over time its rate of expansion has gradually slowed and its constituent particles have come together to form galaxies, planets . . . and us. Eventually, depending on its material density, it will either fizzle out in a "Big Whimper" or shrink back down to an infinitesimal point in a "Big Crunch." These options were delineated by what are technically known as the Friedmann models: simple solutions discovered by Russian cosmologist Alexander Friedmann of the gravitational equations developed by Albert Einstein. The task of cosmology appeared relatively straightforward—to establish the precise age of the universe, firm up the sequence of events and reduce the possible endgames down to one.

Sure there were open questions, but mainstream cosmologists saw these as refinements. Most researchers believed in a clear-cut model of the universe that had little room for change after the first few moments of its history. Much debate was centered on pinning down what happened during the initial ticks of the cosmic clock.

A few of us pondered alternatives to the canon—theories of the universe that strayed from the simplest version of the Big Bang. Like the standard model, these were legitimate mathematical solutions—albeit of variations, reinterpretations, or extensions—of Einstein's equations. Mainstream cosmologists knew about such alternatives but tended to treat them as mere curiosities. In the absence of evidence to the contrary, these researchers advised, why reach beyond conventional approaches?

The situation was akin, in some ways, to the state of affairs before the age of Johannes Kepler and Galileo Galilei. From the 2nd century until the 16th century AD, astronomy relied on the coarse measurements of planetary motion recorded by the Alexandrian thinker Claudius Ptolemy (born circa 85 AD). In his pivotal text, the *Almagest*, Ptolemy developed a clockwork model of the solar system that corresponded well to his data. Consisting of wheels within wheels ultimately turning around Earth, Ptolemy's model showed how planets could follow distinct patterns as they moved across the sky. Because his scheme explained all known facts and fit in well with religious views, scholars found little reason to dispute it. True, it could be simplified, as the Polish astronomer Nicholas Copernicus pointed out, by placing the Sun at the center instead of Earth. But even Copernicus had no new data to back up his proposition.

What changed matters at the turn of the 17th century—as well as at the turn of the 21st century—were substantial improvements in astronomical measuring techniques that led to an enhanced understanding of the movements of celestial bodies. Superior naked-eye measurements of the Martian orbit taken by the Danish astronomer Tycho Brahe led Kepler to conclude in 1609 that the planets follow elliptical paths around the Sun. At approximately that time, images from the first astronomical telescope inspired Galileo to propose that the planets are worlds in their own right and that the stars are distant suns. These findings, in turn, led to the Newtonian portrait of a vast, possibly infinite, universe—home to myriad celestial objects interacting with one another according to the law of gravity.

Telescopes became larger and larger, revealing deeper layers of cosmic order. As they demonstrated, in the race across the celestial plains, stars are hardly lone rangers. Rather, they ride like horses on grand merry-go-rounds called galaxies. Galaxies belong to clusters—assembled, in turn, into even greater superclusters. In 1929, Edwin Hubble, using a colossal device on Mount Wilson in California, discovered that all distant galaxies are receding from each other. This finding led to the standard Big Bang model of an expanding universe—the crown jewel of 20th-century cosmology.

Just as it was enhanced observations that led science to abandon the Ptolemaic model and usher in the modern age, it is the dramatically improved equipment and techniques that have resulted in a rethinking of the standard cosmological approach. In the 1990s and early 2000s astronomy leapt above the clouds with extraordinary new orbiting instruments. Circling high above Earth at distances ranging from hundreds to hundreds of thousands of miles, these telescopic satellites have spanned the spectrum with their light-gathering power. Joining the Hubble Space Telescope, equipped to collect optical light, are infrared instruments, X-ray probes, and several microwave detectors—including the Cosmic Background Explorer and, most recently, the Wilkinson Microwave Anisotropy Probe (WMAP). WMAP has yielded the most precise estimate to date for the age of the universe: 13.7 billion years.

Space-based imaging has been accompanied by other astronomical breakthroughs. Digital cameras, able to absorb and record every single photon (particle of light) streaming down from space, have led to unprecedented precision and deeper-than-ever sky surveys. With these electronic spectacles, once-faint blurs have revealed themselves as extremely distant galaxies that can be analyzed and cataloged. Masterful computer algorithms piece together terabytes of photonic information into detailed three-dimensional images of space. Consequently, for the very first time, astronomy has added realistic depth to its spatial maps.

A leading ground-based project, called the Sloan Digital Sky Survey, has employed these state-of-the-art techniques in a comprehensive three-dimensional scan of a large portion of the northern sky. Mapping more than 200,000 galaxies, the survey has dramatically increased our knowledge of vast segments of space.

Paradoxically, though these instruments and programs have provided more information about the universe than ever before in scientific history, they have revealed how much we really do not know. In particular, they have confirmed a gnawing suspicion among cosmologists that the vast majority of the universe is composed of invisible materials and unidentified energies. As the telescopic results have indicated, only a small fraction of the mass of the cosmos constitutes ordinary matter. The rest is terra incognita! Not only do unseen powers appear to dominate space, they seem to govern its overall dynamics—causing the universe to expand at an ever-increasing rate. In short, we appear to live in an accelerating universe fueled by a hidden dynamo of mysterious origin.

This extraordinary discovery sent shock waves through the world of cosmology, displacing a number of long-held conceptions. No longer can cosmologists focus on the simplest models with the most basic kinds of matter—the textbook examples of expanding universes. Rather, the new findings have revealed more unusual possibilities and solutions.

Some of these novel proposals hypothesize strange new substances with properties unlike anything ever seen. Could, for instance, objects exist with negative mass? Could there be shadow worlds able to communicate with us only through the pull of gravity?

Could there be particles so energetic they have yet to be produced in our particle accelerators? Perhaps the next generation of powerful detectors will reveal such unusual entities.

Other revolutionary schemes involve modifying the law of gravity itself. Could it be that both Newton and Einstein—the greatest geniuses in physics—were wrong about the nature of the gravitational force? Perhaps their portraits of gravity, like unfinished masterpieces, require extra flourishes.

Yet another option involves transforming one or more of nature's constants into a variable. A group of physicists recently speculated that the speed of light could vary over time. Other "variable constant" ideas involve slowly changing values of the fine-structure constant (the parameter governing the strength of electromagnetic interactions), the gravitational constant, and even mass itself.

Finally, some of the most promising approaches for explaining the cosmological mysteries postulate the existence of a fifth dimension beyond ordinary space and time. The fifth dimension arises as a means of unifying all known forces of nature into a single theory. Although its origins date back to the early days of Einstein's general theory of relativity, it has recently been revived in methods for unification called supergravity, string theory, and M-theory.

Traditionally, if a fifth dimension exists, physicists have imagined it to be so small that it could hardly be detected. However, many contemporary approaches envision a large extra dimension, one comparable in scale to conventional space and time. In such a case the fifth dimension could influence the dynamics of the universe and possibly explain why it is accelerating. Moreover, if celestial mechanics is truly five dimensional, the Big Bang need not have been the beginning of time. Rather, it could have been a transition between different cosmic eras. Perhaps the actual cosmos is eternal and its finite age only an illusion wrought by the limitations of our senses.

Indeed, even with the best of all possible astronomical devices, much about the universe could well remain mysterious. Our place in the cosmos is incomparably small; our time in it is but a humming-bird's beat. It would not be too surprising if there are aspects of reality for which we, like dwellers on a tiny desert island, have little knowledge.

One of the greatest learning tools at our disposal is human intuition. Given our peripheral position in the oceans of space, we can use the power of logical deduction to infer much about what lies on distant shores. An outstanding example of the use of human intuition to extend our knowledge far beyond Earth involves the mystery of why the sky is dark at night. By applying some thought to this riddle, there is much we can learn about the universe at large.

1 To See the World in a Grain of Sand: What We Can Observe from Earth

To see the world in a grain of sand
And heaven in a wildflower
Hold infinity in the palm of your hand
And eternity in an hour.

William Blake

IN THE DARK OF NIGHT

It is a familiar stillness—a lull in the rhythm of each day. As the Sun bursts into spectacle and recedes from the sky, ever-darkening shades of color mark its descent. Once the principal player exits the stage, the deep blue backdrop gradually fades into black. Soon no trace of light can be seen, save perhaps the soft glow of the Moon and the pinpoint patterns of the stars and planets.

Most of us take the darkness of the night sky for granted. Yet if one gives it some thought, the blackened nocturnal visage that inspires serenity and sleepiness ought to be a flood-lit, insomnia-inducing glare. Given the vast energies of the universe, pouring radiation ceaselessly down on Earth, we should not need street lamps to navigate nor indoor lighting to read. Evening sporting events ought to be as vivid as day games, midnight strolls as bright as noontime walks.

Think about all the luminous energy constantly bathing our planet. Space contains billions of galaxies, spread out uniformly through the sky. Journeying in any direction, the farther out you go, the more galaxies you'd encounter—passing one after another like mile markers on a Nebraska highway. And a typical galaxy pumps out light at a colossal rate. All these sources of illumination, added up, should be enough to rival sunlight and put an end to the dimness when the Sun isn't even in view. Then why is endless space and, therefore, the night sky not ablaze with light?

The dark sky riddle slipped into astronomical folklore thanks to Heinrich Wilhelm Olbers, a German physician and amateur astronomer. In 1826 he argued that in a uniform, infinite universe populated by eternal and unchanging stars we could potentially see an arbitrarily large number of them. The farther away we looked, the more and more we'd see, because the number would increase as the cube of the distance.

This cubic relationship is like baking larger and larger blueberry muffins. While a thimble-sized pastry might be large enough to contain one blueberry, double each of its dimensions, and it might easily accommodate eight. Make it 10 times bigger in diameter and height and it might even pack in a thousand such morsels. Imagine all the juice that would leak out if all of these were to burst while baking. Similarly, picture all the light produced by greater and greater scopes of stars, each shining in all directions. The amount of illumination heading toward us would be tremendous.

Given a vast-enough cosmos, every single point on the sky should glow with the light of a brilliant star. Bombarded with the colossal radiation of myriad luminous objects, we should need to wear sunglasses night and day. The fact that this does not happen and that the nocturnal sky looks dark has become known as Olbers' paradox.

Scientists wrestled for more than a century and a quarter with this dilemma before beginning to zero in on two conceivable solutions. One possibility, promoted in the 1950s and 1960s by astronomer Hermann Bondi and others, concerns the role of the expansion of the universe in diluting light. Discovered by Edwin Hubble in the 1920s, this expansion reveals itself in the recession (outward movement) of remote galaxies, relative to our galaxy, the Milky Way. Hubble found that the farther away a galaxy, the faster it appeared to be moving away from us. Noting this universal relationship, known as Hubble's law, Bondi argued that the light from distant bodies would become weaker and less energetic the farther it has to travel toward Earth. This effect would greatly dilute the amount of radiation that reaches us, thereby enabling darkness at night.

Considering its critical scientific importance, it was natural for scientists to bring Hubble's law into the discussion. Hubble's revelation of the universal expansion of space is one of the greatest astronomical discoveries of all time. He discovered this effect through a careful study of the atomic spectral lines found in galactic light.

Spectral lines are the fingerprints of atoms, uniquely characterizing their internal structures. As quantum physics tells us, each type of atom has a particular arrangement of energy levels that its electrons can occupy. Like workers ascending or descending a ladder, stepping up and down on certain rungs, electrons are restricted to specific energy states. Each time an electron drops from a higher to a lower level it emits a photon that carries away that energy difference. On the other hand, whenever an electron rises from a lower to a higher level, it must absorb a photon that infuses the required amount of energy. Quantum physics further informs us that photons have wavelike characteristics, vibrating at various rates depending on their energies. The more energetic a photon, the greater its frequency (rate of vibration). Therefore, the energy profile of an atom translates into a unique arrangement of frequencies of the light emitted or absorbed. Physicists refer to these, respectively, as the emission and absorption spectra.

A well-known property of waves, known as the Doppler effect, is that their observed frequencies shift with the speed of the source. A wave moving away from an observer takes on an extra time lag, which makes it appear to vibrate slower. Conversely, a wave moving closer saves time due to its forward motion and seems to vibrate quicker. In the former case its frequency shifts downward, and in the latter case its frequency shifts upward.

Imagine a steady letter writer, traveling around the world, who mails a postcard to a close friend every single day. If she is traveling away from her friend, to lands increasingly remote, her postcards would likely take longer and longer to arrive. Thus, the frequency by which her friend would receive them would steadily drop. On the other hand, if she is making her way back home, her friend's letter box would likely fill up at an increasing pace. Letters sent weeks before from faraway lands might arrive at the same time as those sent days before from nearer locales, leading to a glut of mail. Similarly, as the Doppler effect informs us, the direction of a signal's sender affects its frequency upon arrival.

For sound waves the Doppler effect explains the high-pitched shrieking of a fire engine as it races toward a scene and the low-pitched moan as it speeds away. In the case of light waves, the Doppler effect is visual. Applied to the inward or outward motion of a source, it predicts a shift in luminous frequencies toward the higher or lower ends of the spectrum, respectively. In terms of colors, blue has a relatively high frequency and red a low frequency. Therefore, the increase in frequency for approaching sources (from green to blue, for example) is known as a "blueshift," and the decrease of frequency for receding sources (from orange to red, for instance) is called a "redshift."

Hubble pioneered the use of this effect to probe galactic motions. Pointing the Hooker Telescope (at that time the largest in the world) at various galaxies, he recorded shifts in the frequencies of their atomic spectral lines. He used this information to calculate the velocities (either incoming or outgoing) of each galaxy relative to Earth. Plotting these with respect to galactic distances, he discovered, to his amazement, an unmistakable pattern. With the notable exception of our nearest neighbors (such as Andromeda), all other galaxies in space emit redshifted light—and are therefore racing away from us, like engines from a firehouse. As Hubble observed, the more distant the galaxy, the faster its recessional speed.

The Milky Way isn't alone in being shunned by remote galaxies. We occupy no special place in the cosmos and must assume that the behavior of the galaxies in our region is essentially the same as galaxies everywhere. Therefore, all the distant galaxies must be moving away from each other as well, pointing to a grand expansion of space itself.

Note that the universal expansion does not cause Earth itself to grow bigger. Nor does it cause the solar system to enlarge. Rather, it operates solely on the grandest level: the arena of the Milky Way and its many galactic cousins. Their colossal feud has little effect on our planet—except for telltale signs in the light that the distant galaxies produce. As physicist Richard Price of the University of Texas at Brownsville recently said, "Your waistline may be spreading, but you can't blame it on the expansion of the universe."

The Hubble expansion includes two effects that bear on Olbers' paradox and they concern the energy and density of radiation. The redshifting of light causes it to lose energy. Red starlight, for example, is cooler than yellow. Consequently, as the universe grows, its radiation becomes less powerful. Furthermore, the enlargement of space offers ever-increasing room for photons (light particles). As time goes on, each cubic foot contains, on average, fewer and fewer of these particles. Thus, the Hubble expansion has a double-barreled effect: It cools and dilutes the light in the universe. It makes Earth's night sky darker than it would have been otherwise. Therefore, according to this explanation, the reason we aren't immersed in light is similar to someone trying to take a hot bath in an ever-expanding bathtub. The growth of the bathtub would continuously cool the water and lower its level. Over time, nothing would remain but cold, isolated droplets. By analogy, Earth's night sky displays cooled-down, scattered points of light rather than a warm, luminous flood.

This is an elegant explanation, no doubt. But is it the truth? Sometimes nature baffles us with competing ways of explaining the same effect. The extinction of the dinosaurs, the origin of life, the birth of consciousness, and many other scientific quandaries have triggered formidable debate—with vying accounts struggling for prominence over the years. In this case, science has offered an alternative resolution of Olbers' paradox—one that is different from the Hubble expansion.

Poe's Eureka Moment

Curiously, the true solution to Olbers' paradox has a long literary history. In 1848, Edgar Allen Poe published *Eureka: A Prose Poem*, a volume of his assorted musings about the universe. Recognizing the dilemma of nocturnal darkness, Poe suggested resolution by assuming that light from only a *finite* set of stars has reached us. As Poe wrote:

> Were the succession of stars endless, then the background of the sky would present us a uniform luminosity, like that displayed by the Galaxy—since there could be absolutely no point, in all that background, at which would not exist a star. The only mode, therefore, in which, under such a state of affairs, we could comprehend the *voids* which our telescopes find in innumerable directions, would be by supposing the distance of the invisible background so immense that no ray from it has yet been able to reach us at all.

In other words, Poe divided the universe into two categories. The first part—only a minute fraction—are the stars close enough for their light to have already reached us. The second region—the majority, by far—consists of unimaginably distant objects emitting rays that have yet to touch Earth's skies.

If the universe were infinitely old, Poe's argument wouldn't hold water. No matter how remote a luminous object, we'd witness its light given off some time in its past. For example, if it were billions of light-years away, we'd see its light rays emitted billions of years ago. If it were trillions of light-years away, we'd view its multi-trillion-year-old illumination. This follows from the definition of a light-year: the distance light travels in one year (about 6 trillion miles).

Poe believed and modern science has confirmed, however, that the universe has a finite age. According to current understanding, about 13.7 billion years ago the entirety of space emerged from nothingness (or a prior form) in a mammoth outpouring of energy known as the Big Bang. This material coalesced into whirling galaxies that, in turn, provided the nurseries for myriad stars. Thus, because children cannot be older than their parents, no star has been emitting light for longer than the age of the cosmos. Most stars—the Sun being a good example—are in fact much younger.

A key prediction of the Big Bang theory was confirmed in 1965 through the detection of radiation left over from its initial stages. The relative uniformity of this relic radiation indicated that it had a common source—namely, the explosive beginning of the universe. Thus, the direct cause of the Hubble expansion was the emergence of the cosmos from a fireball.

The same year as the Big Bang's verification, University of Massachusetts astronomer Edward Harrison made use of this finding to spin a modern version of Poe's tale. Because all the stars and galaxies have finite ages, he noted, there has not been enough time for space to flood with their light. Therefore, when we gaze at the night sky we see only the illumination of objects that are close enough and old enough for their light to have reached us. This illumination represents only a small amount of light, leaving the rest of the heavens as black as the depths of a cave.

This situation can be compared to the maximum range of signals from television stations on Earth. One of the first experimental broadcasts took place in the late 1920s, transmitting an image of the cartoon character Felix the Cat. Traveling at the speed of light, by now Felix's visage has spread out across a spherical region of space almost 80 light-years in radius, encompassing hundreds of stars. Any intelligent beings residing within that shell, capable of interpreting our signals, would know that we have television (and would conceivably become acquainted with Felix). Yet because television is relatively new, the vast bulk of the universe has yet to encounter our broadcasts. Thus, the fraction of the cosmos containing our television signals is virtually nil. Similarly, because of the finite age of stellar "broadcasts," the intensity of local starlight is extremely small.

For several decades, both explanations for Olbers' paradox appeared in various texts. At one point—by our estimate—about 20 percent of the books said the night sky is dark because the galaxies are receding from each other, while another 30 percent said the reason lies in their finite age. The rest mentioned both factors but did not say which is more important. The lack of a clear answer to such a fundamental question seemed a scandal of the first order.

In the mid-1980s, one of us (Wesson) along with two colleagues, Knut Valle and Rolf Stabell of the University of Oslo, set out to nail down this matter in a way the entire astronomical community would accept. A series of fruitful discussions led to the definitive resolution of the problem. The results were published in the June 15, 1987, issue of the *Astrophysical Journal,* the leading journal in its field.

To construct its solution, the team conducted a clever thought experiment. In its alternative version of cosmic reality, it pictured preventing the expansion of the universe while keeping all other properties of the galaxies (particularly their ages) the same. In this manner the group ascertained the intensity (brightness) of intergalactic light in a static universe, implying that the finite age of the galaxies was the determining factor. Then it allowed the expansion to resume, knowing that both the age factor and the expansion factor controlled the intensity of intergalactic light. Finally, it found the ratio of these intensities, establishing the relative importance of each factor.

The group's experiment was like observing a city scene during moonless and moonlit nights to see whether moonlight or street lamps contributed more to urban illumination. By varying one feature while maintaining the others, it was straightforward to compare their effects. The results were undeniable, setting aside years of controversy. The team estimated the typical ratio of intergalactic light with expansion to be only about one-half of that without it. Given the flood of energy produced by all the galaxies, the expansion of the universe thereby made little difference. Rather, the age factor produced by far the greatest impact, robbing the heavens of the vast majority of its potential brightness. Hence, even in a static universe, the sky would be plenty dark.

This experiment resolved Olbers' paradox once and for all: The night sky is dark because the universe is still young, not because it is expanding. So the next time you stub your toe in the dark of night, you can justifiably blame it on the Big Bang.

WHERE ARE THE ALIENS?

Not only is space black, it is also silent. Surrounded by endless stellar reaches, Earth seems a lonely outpost devoid of communications from any other world. Given the likelihood of planetary systems orbiting many (if not most) of the stars we see, why hasn't a single one of them sent us a simple hello? Could space be as empty of life as it is of light, or might there be another explanation?

The great Italian physicist Enrico Fermi pondered this dilemma in 1950 while taking a break from the rigors of Los Alamos Laboratory. It was the era of the UFO craze, and newspapers were brimming with speculations about the prospects for alien visitors. A clever cartoon about the subject caught his eye and led him to estimate the probability of extraterrestrial contact. During a casual lunch, he raised the topic with three of his colleagues—Edward Teller, Herbert York, and Emil Konopinski. While discussing sundry matters, Fermi suddenly asked, "Where is everybody?"

Fermi's lunchtime companions knew him as a man of deep thought and did not take his question lightly. As an expert in seat-of-the-pants calculations, he was adept at ruling in or out various physical scenarios. If the esteemed planner of the first self-sustained nuclear reaction was troubled by a missing element, chances were that something was wrong.

The "everybody" in question referred to the preponderance of extraterrestrials our vast universe ought to contain. As Fermi pointed out, given a sufficient number of worlds in space, at least a fraction of them should harbor civilizations advanced enough to attempt contact with us. Then, considering that the cosmos has been around for billions of years, why haven't any of them sent signals by now? The curious situation that Earth has never encountered alien communications has come to be known as Fermi's paradox.

Belief in the abundance of life in the universe dates at least as far back as the early days of the scientific revolution. Many notable thinkers have stressed that if life emerged from Earth's once-barren soil, it ought to have arisen on countless other planets as well. Isaac Newton, for example, once wrote, "If all places to which we have access are filled with living creatures, why should all these immense spaces of the heavens above the clouds be incapable of inhabitants?"

Ten years after Fermi's remark, astronomer Frank Drake developed an equation that has come to epitomize the chances of encountering intelligent life in the universe. The Drake equation consists of a number of multiplicative factors, each of which represents one aspect of the likelihood for alien contact. These factors include $N*$, the number of stars in the Milky Way; f_p, the fraction of stars that have planetary systems; n_e, the average number of planets in each system that have environmental conditions suitable for life; and f_l, the chances that life actually arises. The final three parameters pertain to the emergence of intelligence itself: f_i, the fraction of life-nurturing worlds with intelligent beings; f_c, the fraction of those with cultures advanced enough to broadcast messages; and f_L, the longevity of such civilizations. Except for $N*$ and n_e, each of these factors ranges from 0 to 1 (with 0 representing "none" and 1 representing "all"). The product of all these factors yields an estimate of

N, the number of civilizations in our galaxy potentially able to contact us. In equation form, Drake wrote this as:

$$N = N* f_p \, n_e \, f_l \, f_i \, f_c \, f_L$$

Some of these factors are more quantifiable than others. For example, models of stellar and planetary formation are fairly well developed. Although at the time Drake proposed his equation scientists knew of no other planetary systems, in recent years they have discovered more than 100 worlds beyond the solar system. As a consequence of these findings, researchers have developed superior estimates for the fraction of stars with planetary companions. Astronomer Geoff Marcy, one of the leading planet hunters, has recently surmised that roughly half of all stars have planetary systems.

Scientists have been far more tentative about the prospects for life and intelligence on other worlds. Because no living organisms have been found yet in space, let alone cognizant beings, the components of the Drake equation pertaining to these possibilities are still highly speculative. Nevertheless, Drake and other astronomers have offered guesses as to their ranges. Carl Sagan, for example, argued that the number of advanced civilizations capable of communicating with us could be as low as 10 or as high as in the millions, depending on their capacity to avoid nuclear destruction.

Drake and Sagan were leading proponents of the Search for Extraterrestrial Intelligence (SETI), a systematic hunt for radio signals from alien civilizations. Beginning in the 1960s, radio dishes around the world have scanned the skies for telltale coded patterns. In the intervening decades the SETI program has been greatly expanded, encompassing a wider range of frequencies and a broader array of targets. Improved software and faster processing rates have made it easier to wade through the haystack of radio noise, thereby enhancing the prospects for uncovering buried messages. Alas, despite a number of false alarms, not one has been found.

In the 1970s, baffled by the constant lack of evidence for extra-terrestrial civilizations, a number of scientists put forth proposals suggesting that advanced life in the cosmos (or at least our galaxy) is extremely rare. The most famous of these proposals was a paper by astronomer Michael Hart, advancing the startling proposition that we are the first civilization in the Milky Way. Hart reached this con-clusion through a systematic study of conditions that could realisti-cally affect alien communication, none of which would present enough of an obstacle to persistent aliens who wanted to contact us. If extraterrestrials that were currently capable of radio transmissions existed, we surely would have heard from them by now. Thus, Hart answered Fermi's famous question with the discouraging solution that nobody able to talk to us is around yet.

Tulane physicist Frank Tipler amplified Hart's suggestion with a detailed explanation for ruling out the existence of intelligent extra-terrestrials beyond Earth. Tipler argued that any extant advanced civilization would have at some point in its history developed the means for galactic colonization through an army of self-reproducing robot ships. These automata would be programmed to explore new worlds, establish outposts on the most suitable ones, and then fashion replicas of themselves to repeat the process elsewhere. By now, Tipler contended, the Milky Way would be replete with signs of one or more such civilizations. The absence of such signs led Tipler to con-clude that the only intelligent beings in our galaxy were his fellow humans.

BEYOND THE COSMIC HORIZON

Even if advanced life is rare in the Milky Way, that does not preclude an abundance of civilizations in other galaxies. An infinite universe would render even the slimmest chance for intelligence a reality *some-where* else in space. Given enough room in the cosmos and enough time for intelligence to develop, the cosmic roulette wheel would be bound to hit the lucky number. It would be just like placing a million monkeys in front of a million computers and letting them bang on the keyboards for an extremely long time. Eventually, through their random actions, one of them would type a Shakespearian sonnet.

The lower the probability for intelligent life to evolve, the farther we need to look to find it. Hence, before drawing conclusions about the current failure of the SETI mission to discern signals from within the Milky Way, we must expand our search to include other galaxies. Although the present-day program envisions civilizations with the capacity to broadcast messages over tens or hundreds of light-years, we can easily imagine extragalactic cultures with even greater capabilities. Moreover, because each galaxy potentially harbors hundreds of billions of worlds, there could very well be far more civilizations able to reach us with their signals outside the Milky Way than within it. Therefore, by aiming our radio dishes at intergalactic as well as intragalactic targets we might improve our search for extraterrestrial intelligence.

Suppose comprehensive scans of the heavens—including the broadest possible scope of galaxies—continue to fail to turn up signs of sentient life. Should the science community then conclude, like Tipler, that we are alone in the cosmos? Or could there be another reasonable explanation for the complete lack of communication?

To address this issue, let's draw a valuable lesson from the way we resolved Olbers' paradox. In that case we found that the finite age of galaxies and the finite speed of light conspire to shield us from the totality of radiation emitted in space, letting only a minute portion reach our skies. Similarly, perhaps the finite age of extraterrestrial civilizations and the finite speed of light preclude us from receiving alien broadcasts. In contrast to starlight and galactic light, maybe the effect is so severe that not even a single message would be able to reach us.

If that seems odd, think about the case of the Felix the Cat signals broadcast in the 1920s. Because of the limitations posed by the speed of light, only a small fraction of the stars in the galaxy are close enough (within 80 light-years) to have already come into contact with those signals. Suppose intelligent life is rare enough that none of these nearby stars have planets inhabited by civilizations capable of radio communication. Then presently no other world in space could possibly know about Felix and his human creators.

If we now imagine intelligent life so uncommon that the nearest communicative civilizations lie in remote galaxies, we can see how vast distances could preclude contact. A culture millions or billions of light-years away would have had to be broadcasting for eons in order for us to know about them. If we use the history of life on Earth as the model, many planets would have been too primitive to support advanced beings that long ago. Therefore, no communicative civilization would be ancient enough for its signals to have already reached us.

Moreover, on galactic scales the expansion of the universe would greatly exacerbate the time delay. Because the alien races would be situated in galaxies fleeing from ours, their radio broadcasts would need to cross ever-widening gulfs. Hindered by the currents of outward galactic movement, any messages sent out would wash up on our shores far, far later than they otherwise would in a static universe.

If the closest civilizations are even farther away, we would never learn of their existence. As cosmology tells us, beyond an invisible barrier called the *particle horizon*—defined as the greatest distance any incoming particle could have traversed in the universe's current age—alien signals wouldn't stand a chance of reaching us. They'd face the situation of Alice in the looking-glass world; though they'd travel as fast as they could, they wouldn't be able to outrace the expansion of the universe.

Indeed, it's entirely possible that a cornucopia of worlds could reside beyond the curtain of invisibility. Some of these planets might even be Earth's near twin. Others could house technologies exceeding our wildest speculations. Yet unless any of these societies find a way of circumventing the speed of light, we would remain as separate from them as prisoners confined forever to solitary cells.

From our resolutions of the paradoxes posed by Olbers and Fermi we have seen that there really isn't just one universe. The *observable* universe—consisting of all the galaxies within communication range—comprises but a fraction of the entire physical universe. Features of the former, including the relatively sparse nighttime sky and the lack of evidence for extraterrestrial signals, do not necessarily reflect the complete cosmos. We could well be living in an infinite space with unlimited sources of energy and myriad worlds wholly beyond our perception.

THE FRONTIERS OF KNOWLEDGE

An integral part of the human condition is that we are faced with limits. Our senses and abilities can take us only so far. Beyond them lie vast stretches of unknown territory. None of us have been to the surface of Pluto or to the bottom of the Marianas trench. No one has ventured into the center of the Earth or traveled backward through time. Confined through mortality to just a tiny sliver of eternity, we will never experience the distant past or the far future. Yet we are a race of dreamers and cannot help but ponder the wonders that might reside in places beyond our reach.

Our intellect yearns for knowledge of the cosmos in its entirety. Therefore, it is frustrating to think that much, if not the bulk, of the physical universe lies outside our range of observation. It is even more unnerving to realize, as recent evidence has shown, that the conventional material we *do* detect is far outweighed by invisible material. Atoms and molecules—the stuff of planets and stars—are but minority occupants of space. The major players are bizarre entities known as dark matter and dark energy. Until we fathom these substances, we have taken only a child's step toward comprehending the universe as a whole.

Dark matter was originally postulated as an explanation for unexpected discrepancies between the actual and predicted motions of certain celestial bodies. In the early 1930s, Dutch astronomer Jan Oort noticed that the stars in the Milky Way tend to be drawn much more tightly to its central plane than Newton's law of gravitation would require. Estimating the theoretical value of the collective pull of our galaxy on each of its stars, Oort found that the observed amount is three times greater than expected. It is like a ghostly tug-of-war with powerful apparitions assisting each live player.

Shortly thereafter, Swiss physicist Fritz Zwicky discovered a related effect concerning the behavior of galaxies in clusters. A cluster is a stable collection of galaxies, held together through the force of gravity. Examining the Coma Cluster (in the constellation Coma Berenices), Zwicky calculated the amount of matter needed to provide its gravitational "glue." To his astonishment, he found that the required mass is hundreds of times what astronomers observe telescopically. (He had made an error in his assumptions, but even with the correction there was a significant discrepancy.) Zwicky postulated that the bulk of the material in the Coma Cluster is invisible.

It wasn't until the late 20th century, however, that the mainstream scientific community reached the unmistakable conclusion that there is far more to the cosmos than meets the eye. Astronomical sleuthing, gleaning results from a phenomenon known as gravitational lensing, demonstrated that dark matter pervades the universe—from the hearts of galaxies to the voids of deepest space. This method measures the bending of light from distant objects due to the gravitational influence of intervening bodies—seen or unseen. It relies on a concept proposed by Einstein in his general theory of relativity—massive objects warp the fabric of space and time, causing photons and other particles to alter their paths. Thus, the weighty presence of matter—even invisible material—can bend light like a lens. Astronomers can determine the amount of distortion in a section of space by observing changes in the apparent brightness or position of the light sources passing behind it—like watching bugs alter in appearance as they crawl beneath a magnifying glass. Then these observers can calculate how much mass must have caused the curving. Offering an extraordinary tool for mapping out the hidden material in the cosmos, gravitational lensing has furnished ample evidence that luminous bodies comprise just a small subset of all that there is.

What is this mysterious substance that signs its name only with gravity's mark? Early on, scientists supposed that it was nonshining stars, meaning those that either burned out or never had enough material to ignite in the first place. Examples of these would be objects called neutron stars (the ultracompact remnants of massive stellar cores) and brown dwarfs (stars comparable in size to very large planets, lacking the critical mass of hydrogen required to stoke the furnace of stellar fusion). Further measurements, though, have indicated that nonshining stars represent only a portion of the missing material. Most of the hidden stuff must be composed of new kinds of substances—rather than the ordinary matter, made of protons and neutrons, that constitutes stars and planets.

Cosmologist Michael Turner of the University of Chicago has offered a number of suggestions for what dark matter could be. At scientific conferences he lays out his ideas on colorful transparencies, wagering like a sports commentator which are the best bets. His prime candidate is a hypothetical particle called the axion, to which he gives high odds despite the fact that powerful detectors have searched for it in vain.

To further complicate matters, in 1998 an extraordinary astronomical discovery seemed to cast even more of the cosmos into shadow. Using precise measurements of the distances and velocities of supernovas (stellar explosions) in extremely remote galaxies, several teams of astronomers determined how the Hubble expansion changes over time. To their amazement, they found that the universe's growth is speeding up as it ages. Not only is the cosmos ballooning outward, it is doing so faster and faster—with no end in sight.

Until the supernova findings, many astronomers assumed that the long-term evolution of space would constitute one of two possibilities depending on the amount of mass within it: either continuing to expand forever at a slower and slower pace, slowed by the mutual gravitational attraction of all its matter and energy, or, if its density exceeds a certain critical amount, reversing course and recontracting down to a crunch. The options resembled a roller coaster nearing the end of its track. Virtually everyone expected a gradual slowdown, followed perhaps by a backward ride. Few thought the vehicle would be charging full speed ahead.

Ordinary gravity cannot account for such acceleration. As an attractive force, it acts to clump massive objects together, putting brakes on the outward motion of the galaxies. Because both dark and visible forms of matter interact on the basis of gravitation, they could not engender the repulsive forces required to push galaxies apart. Turner and other researchers rapidly reached the conclusion that a new type of substance must be at work, one that creates a kind of cosmological antigravity. They dubbed the unknown agent "dark energy" to distinguish it from dark matter.

There are several important differences between dark matter and dark energy. While dark matter is thought to have an uneven distribution, mainly clumped around visible population centers (with a lesser amount sprinkled throughout the void), dark energy is believed to be as smooth as custard, spread uniformly throughout space. Otherwise, in contrast to known observations, the universal expansion would exhibit distinct behavior in various directions.

Moreover, although the composition of dark matter is largely unknown, scientists have put forth an array of likely candidates. Any massive, but elusive, particle present in sufficient quantities could potentially fit the bill. A prime example is the neutrino, a fast-moving particle believed to comprise at least a portion of dark matter.

By comparison, dark energy contenders have been much harder to identify. Proposed explanations have called on entirely new physical paradigms, stretching the limits of our imagination. Some of the suggestions include introducing an altogether novel type of energy field called quintessence or even modifying the law of gravity itself.

Recent results from probes of the cosmic microwave background have pinned down the relative abundance of dark matter and dark energy on the one hand, compared to luminous materials on the other. According to the WMAP survey, considered the most precise scan of cosmic background radiation ever conducted, about 23 percent of the mass of the observable universe is composed of dark matter, about 73 percent is dark energy, and only 4 percent is ordinary visible material.

Presented with such startling evidence of our minority status in a vast and dark cosmos, scientists can no longer assert that their celestial charts reflect the true picture of reality. These findings have presented cosmology with one of its greatest challenges in history: shedding light on the shadowy substances that dominate the physical world.

CALLING ON EINSTEIN

Intuition took us far in pondering solutions for Olbers' paradox and the Fermi paradox. By pressing forth the ramifications of several basic principles—the finiteness of the speed of light, the limited age of the universe, and the Hubble expansion of space—we found ways to explain nocturnal darkness and the lack of alien communication in the face of a possibly infinite universe. Keeping these successes in mind, let's apply scientific reasoning to cosmology's greatest enigmas—including the puzzles of dark matter and dark energy.

One of the great champions of thought experiments, Albert Einstein, developed a remarkable equation that will aid us in our pursuit. It is not his most famous equation, linking energy and mass, but rather a relationship between the matter and geometry of any region of space. The basis of his general theory of relativity, it demonstrates how matter affects the universe itself. Not only does it predict the Hubble expansion, it also yields precise forecasts for what happens if the mixture of dark matter, visible matter, and energy is altered. Moreover, it even includes an antigravity term, called the cosmological constant, that can be interpreted as representing the impact of dark energy on universal dynamics.

The route Einstein took to his grand equation was extraordinary. Putting forth bold insights about gravitation, accelerated motion, and the roles of space and time, he crafted this raw material through the machinery of mathematics into a beautiful edifice unmatched for its elegance and simplicity. Showing little wear for its age—at least until recently—this construction has provided sturdy support for the burgeoning field of cosmology.

Although one is loathe to tamper with success, it could be that Einstein's construct will require reinterpretation or even modification to bear the added weight of contemporary astrophysics. Before considering such options, however, let us retrace Einstein's steps and examine how he assembled various physical suppositions into a masterpiece of mathematical architecture.

2   Infinity in the Palm of Your Hand: Einstein's Far-Reaching Vision

I myself am of Mach's opinion, which can be formulated in the language of the theory of relativity thus: all the masses in the universe determine the [gravitational] field. . . . In my opinion, inertia is in the same sense a communicated mutual action between the masses of the universe.

Albert Einstein (response to Ernst Reichenbaecher)

The fault, dear Brutus, is not in our stars, but in ourselves.

William Shakespeare (*Julius Caesar*)

TOUCHED BY THE STARS

The ancients believed that celestial patterns steered the individual fortunes of human beings and the collective destinies of civilizations. For example, if a particular constellation, or stellar grouping, was high in the sky on the night a certain king was born, he would be blessed with the fiery gifts of a warrior. Another heavenly configuration and he was doomed to die in battle. If Venus kissed Jupiter in the chapel of lights, then a royal marriage was brewing. But if the stars were all wrong on the night of marital bliss, the bride would alas be barren.

Is it lunacy to believe that there is a deep connection between earthly and celestial events? Not if one takes the concept literally.

The term "lunacy" itself derives from beliefs in periodic influences of the Moon. As the shining beacon of the nocturnal sky, Earth's satellite was thought to exert quite a pull on terrestrial affairs.

There is little evidence that the Moon has driven anyone mad, or induced anyone to sprout extra facial hair. Yet, especially for those attuned to the rhythms of the sea, it clearly exerts a pull on many lives. For those who earn their living hauling in lobsters from the Bay of Fundy off the coast of maritime Canada, each working day is shaped by lunar forces. Amid some of the highest tides in the world, one could not help but concede that the sandstone of human destiny is carved by heavenly guided waters.

Today we distinguish between scientific forces and spiritual influences. To the ancients this distinction was not so clear. Early astronomers did double duty, serving both to record the positions of the celestial spheres and to apply this information for astrological forecasts. Their expertise in predicting eclipses, planetary conjunctions, and other celestial events, as well as offering critical navigational knowledge, earned them the mantel of exalted prophets.

Even as late as the 16th century, many scientific researchers, such as the German mathematician Johannes Kepler, sold horoscopes on the side for extra income. Kepler, in his first astrological calendar, proudly predicted a cold spell and a Turkish invasion of Styria (now Austria). Not only did he peddle forecasts, he deeply believed that they offered special insight into the determinants of human character. He once wrote that his father was "vicious, inflexible, quarrelsome and doomed to a bad end" because of the clashing influences of Venus and Mars.

How did the heavenly orbs set the pace of their own motions and influence the course of earthly events? Kepler originally thought this happened because the planets somehow possessed minds of their own. However, after he developed a clearer understanding of celestial mechanics, he realized this could not be the case. "Once I firmly believed that the motive force of a planet was a soul," he wrote. "Yet as I reflected, just as the light of the Sun diminishes in proportion to distance from the Sun, I came to the conclusion that this force must be something substantial."

Thus, what ultimately changed Kepler's opinions on these matters was the realization that planetary motion could be explained through simple mathematical rules. This revelation came through a systematic study of the orbital behavior of the planet Mars and generalizing these results to other bodies in the solar system. Kepler discovered that each planet travels along an elliptical path around the Sun, sweeping out equal areas (of the region inside the ellipse) in equal times. He also found relationships between each planet's orbital period and its average distance from the Sun. These discoveries, known as Kepler's laws, led him to conclude that rock-solid mathematical principles, not ethereal spiritual influences, govern celestial mechanics.

Picking up where Kepler left off, Isaac Newton brilliantly revealed the mainspring of this clockwork cosmos. He discovered that the same force that guides acorns down from oak trees and cannonballs down to their targets similarly steers the Moon around Earth and the planets around the Sun. Calling this force gravitation, from the Latin *gravitas* or heavy, he showed that it exerts an attractive pull between any pair of massive objects in the universe. The Moon, for instance, is pulled toward Earth just like a ripe fruit from its branch. Earth is similarly drawn toward the Moon—which explains the movements of the tides.

Newton further demonstrated that the strength of gravity varies in inverse proportion to the square of the distance between any two masses. That is, if two objects are flung twice as far away from each other, their gravitational attraction drops by a factor of 4. This reduction in strength with distance explains why the Moon, rather than any of the stars (as massive as they are by comparison), lifts and lowers Earth's ocean waters.

Some contemporary believers in astrology have asserted that the marching parade of constellations exerts a changing gravitational influence on the temperaments of children born under these signs.

When disaster strikes they like to think that the fault lies not in ourselves but in our stars. (Indeed, the word "disaster" derives from the Latin for "the unfavorable aspects of stars.") However, as the late astronomer Carl Sagan was fond of pointing out, the gravitational attraction of the delivering obstetrician, hands cupping the baby's head, outweighs the pull of any distant star. Though the stars are far more massive, the obstetrician is much, much closer. Besides, it's unclear how any gravitational force could affect thought processes, unless one hangs upside down to permit greater blood flow to the brain.

When applied to the proper venue, material objects in space, Newtonian theory is remarkably successful. Yet it harbors an essential mystery. Why do bodies orbit their gravitational attractors, rather than moving directly toward them? Why doesn't the Moon, for instance, immediately plunge toward Earth and destroy all civilization?

Newton's answer was to propose a property called inertia that keeps still objects at rest and moving objects traveling in a straight line at a constant speed. Like a universal hypnotist, inertia places each object in a trance to continue doing what it is already doing. The only thing that can break inertia's spell is the application of an external force (or unbalanced set of forces). Still, the magic is lifted only provisionally, allowing the body to change paths only during the interval in which the force is applied. Then it resumes its straight-line motion, until perhaps another force takes hold.

Now consider the case of the Moon. Inertia compels the Moon to keep going in a straight line, but gravity continuously pulls it toward Earth. The compromise is a curving motion, resulting in an essentially circular path.

Though gravity is a force, inertia is not. Rather, inertia represents the state of nature in the *absence* of all forces. As strange as it might seem, according to Newtonian theory, if all the forces in the universe suddenly "turned off," every object would continue moving forever uniformly. This state of affairs would result from no specific cause but rather from a lack of causes.

In trying to fathom the underlying reason for inertia, one is reminded of the Taoist paradox that, in trying to pin down something's definition, its true meaning slips away. The machinery of inertia is remarkable in that there is no machinery. Nevertheless, from Newton's time onward, physicists and philosophers have sought a deeper understanding of why constant linear motion constitutes nature's default mode.

To make matters even trickier, all motion is relative. This principle was put forth by Galileo and firmed up by Newton, well before the time of Einstein. The speed of any object depends on the frame (point of view) in which it is observed. For instance, if two truck drivers, traveling at the same speed but in opposite directions, wave to each other on a highway, each will observe the other to be moving twice as fast as their speedometers would indicate. If, on the other hand, each is traveling at identical speeds in the same direction, each will appear to the other to be at rest—presuming they ignore all background scenery.

You would think that the property of inertia would similarly be relative. If, according to one reference frame, inertial motion appears unblemished by extra forces, why not in all frames? Strangely, though, while this is true for observers moving at constant velocities with respect to each other, it is emphatically not true for accelerating observers. Newton cleverly demonstrated this principle by means of a simple thought experiment involving a spinning bucket of water.

BEYOND THE PAIL

Sometimes the most ordinary household objects can offer deep insights about the physical universe. If we concoct the right experiment, there is no need for an expensive particle accelerator to probe the mysteries of force, nor a high-powered telescope to reveal the enigmas of the cosmos. A visit to our basement or backyard might well provide all the materials required.

Take, for example, an ordinary pail. Fill it to the brim with plain tap water. Suspend the bucket from a rope, attached to the limb of a tree. If the bucket is still, the water should appear completely level.

Not the stuff of Nobel prizes, so far, but here's where things get strange. Spin the bucket. Twirl it around gently but resolutely. As the bucket turns, you notice several things. First, the water remains in the pail. Thanks to inertial tendencies, it pushes against the walls of the bucket but doesn't spill out.

Yet something does change about the fluid. Its surface begins to hollow out, as if sculpted by a potter. In short order, the once-level top has become as curved as a soup bowl.

The principle of inertia can explain this concavity, but only if you adopt the right perspective. From your point of view, the reason is simple: The water is building up against the sides because, despite the spinning of its container, it wants to travel in a straight line. This lowers the central part of the fluid, hollowing it out.

Consider, however, the perspective of a tiny observer (a savvy ant, perhaps) perched on the side of the pail. If he ignores the world beyond the bucket, he might well believe that the bucket isn't spinning at all. For him, therefore, inertia should keep everything inside the bucket at rest. Then, imagine his surprise if he looks down at the water and sees it change shape. What bizarre supernatural effect, he might wonder, could deliver such a targeted punch?

Newton used his bucket argument to make the point that, while the principle of inertia does not depend on the relative *velocity* of two reference frames, it clearly does depend on the relative *acceleration* of the frames. In physics, acceleration refers not just to alterations in speed but also to changes in direction. Therefore, a spinning bucket is accelerating because the motion of any point within it keeps changing direction. But indeed that is true about Earth itself—rotating about its own axis as it revolves around the Sun. Therefore, given all these gyrating vantage points, how can we uniquely define inertia's unmistakable action? Where in this whirling cosmic carnival can we find solid ground?

Newton's answer was to define a fixed, universal reference frame, called "absolute space," placed in an exalted position above any other framework. "Absolute space," he wrote, "in its own nature, without relation to anything else, remains always similar and immovable."

To the concept of absolute space, Newton added another expression, called "absolute time." Absolute time represents the uniform ticking of an ideal universal clock. Together, absolute space and time serve to define absolute motion—an inviolate description of movement through the cosmos.

Common parlance, Newton pointed out, fails to distinguish between relative motion (measured with respect to any fleeting frame) and absolute motion (defined with regard to the steel scaffolds of absolute space and time). The bucket example, however, demonstrates why such confusion of terms won't do. To understand dynamics properly, he emphasized, one must reject the ephemeral and take a firm universal perspective. "Relative quantities," he wrote, "are not the quantities themselves whose names they bear. . . ." Those who mistake transient measures for true quantities, he continued, "violate the accuracy of language, which ought to be kept precise. . . ."

Despite Newton's admonition, in the centuries after his death a growing community of scholars came to find his distinction rather artificial. With everything in the cosmos in ceaseless motion, why should any one reference frame stand still? By the 19th century, a number of scientists replaced Newton's artifice with an all-pervading invisible substance, known as the aether. Absolute motion could thereby be defined with respect to the aether stream. Nobody, however, could detect the aether; it seemed as elusive as a ghost.

Viennese physicist Ernst Mach took a different approach. In his popular book on mechanics, he dismissed the notion of an absolute frame. Rather, he argued that it is the combined pull of distant stars that keeps inertia's hammock aloft. "Instead of referring a moving body to [absolute] space," he wrote, "let us view directly its relation to the bodies of the universe, by which alone such a system of coordinates can be determined." Hence, objects resist acceleration because they are in some mysterious way "connected" to the myriad other bodies in the cosmos. Even a lowly pail on Earth responds to mammoth energies trillions of miles away. This far-reaching concept has come to be known as Mach's principle.

It's strange to think of remote stars steering the water in a bucket. Yet the idea that the Moon guides the tides of Fundy seems perfectly normal. Mach's principle just stretches such cosmic connections much, much further—until they engulf the entirety of space itself. When Mach published his treatise, he freely admitted that he had no experimental proof for his hypothesis. Yet because it was based on the actions of real celestial bodies, he asserted that his explanation was heads above Newton's abstract design. This argument stirred the youthful imagination of Albert Einstein, who dreamed of putting Mach's ideals into practice.

COMPASSES AND CLOCKS

Albert Einstein, the greatest physicist of modern times, was born in Ulm, Germany, on March 14, 1879. As a child he had a keen curiosity about the principles underlying the way things work. In an autobiographical essay, he recalled his wonder at the age of 4 or 5 when he was lying ill in bed and his father presented him with his first compass.

"The fact that the needle behaved in such a definite manner did not fit at all into the pattern of occurrences which had established itself in my subconscious conceptual world (effects being associated with 'contact'). I remember to this day—or I think I remember— the deep and lasting impression this experience made on me. There had to be something behind the objects, something that was hidden."

Then, at the age of 12, a family friend gave him a book on Euclidean geometry. The young thinker marveled at the crisp certainty of the mathematical arguments presented. Soon he learned how to construct his own proofs, creating geometric rules from simple propositions.

Like many philosophers before him, Einstein was intrigued by the contrast between the imperfect arena of sensory experiences and the ideal realm of abstract concepts. He wondered which aspects of the world required hands-on experimentation and which could be deduced through pure thought. His life's journey stepped carefully between these two positions. Ultimately, the latter would win out, and his mathematical side would overtake his more practical side. He would become obsessed by the idea of finding inviolable mathematical principles, elegant and beautiful in their simplicity of expression, that could explain all of nature.

One of Einstein's first "thought experiments" involved a seeming contradiction between Newtonian physics and the known properties of light. At the age of 16 he imagined chasing a light wave and trying to catch up with it. He pictured himself running faster and faster until he precisely matched the speed of the flash. Then, he wondered, would the signal seem still to him, like two trains keeping pace?

Newtonian physics would suggest the affirmative. Any two objects moving at the exact same velocity should observe each other to be at rest—that is, their relative velocity should be zero. However, by Einstein's time, physicists knew that light was an electromagnetic wave. James Clerk Maxwell's well-known equations of electromagnetism made no reference to the velocities of observers. Anyone recording the speed of light (in a vacuum) must measure the same value. Hence, two of the giants of physics, Newton and Maxwell, appeared locked in a conceptual battle.

Others tried to find a way out of this dilemma by proposing effects due to the invisible aether (which by that time had experimentally been discredited), but it was Einstein who developed the definitive solution. In a breakthrough known as the special theory of relativity, he demonstrated that Newtonian mechanics and Maxwellian electrodynamics could be reconciled by abandoning the notions of absolute space and time. By asserting that measured distances and durations depend on the relative velocities of the observer and the observed, Einstein developed dynamical equations that preserve the constancy of the speed of light.

Let's see how special relativity works. Suppose a runner tries to catch up to a light wave. As he moves faster and faster, approaching light speed, his personal clock (as measured by his thoughts, his metabolism, and any timepieces he is wearing) would slow down relative to that of someone standing still. Compared to the tortoise-like ticking of his own pace, light would still seem to be whizzing by at its gazelle-like speed. He wouldn't know that his own time is moving slower, unless he later compares his findings with a stationary observer. Then he would realize that he had experienced fewer minutes while running than he would have just by standing. All this ensures that any observer, whether moving or still, records exactly the same value for light speed.

This phenomenon, of clocks ticking more slowly if they move close to the speed of light, is called time dilation. Time dilation is nature's hedge against anyone catching up with its fastest sprinter. Nature would rather slow down stopwatches than allow runners carrying them to violate its sacred speed limit.

A related mechanism, called length contraction (or sometimes Lorentz-Fitzgerald contraction), involves the shortening of relativistic objects along the direction of their motion. This is a clear consequence of time dilation and the constancy of light. If one uses a light signal to measure a length (by recording how long it takes to go from one end to the other) but one's clock is slower, one would naturally find the object to be shorter.

In 1907, two years after Einstein published his special theory of relativity, the Russian-German mathematician Hermann Minkowski proposed an extraordinary way to render it through pure geometry. Minkowski suggested that Einstein's theory could be expressed more eloquently within a four-dimensional framework. With a thunderous speech, he proclaimed the very end to space and time as separate entities, replacing them with unified four-dimensional space-time.

Within this framework, known as Minkowski space-time, anything that happens in the universe is called an "event." Spilling a morning cup of coffee in a Ganymede café could be one event; shipping out emergency supplies of Venusian organically grown house blend on a sweltering afternoon could be another. The "distance" between these two occurrences, called the space-time interval, involves combining the differences in time and space between the two events.

How, you might ask, can time be "added to" space? The answer involves using nature's universal constant velocity, the speed of light. Multiplying velocity (in miles per hour, for example) by time (in hours) yields length (in miles). The time units cancel out, leaving only length units. Multiplying all time values by the speed of light converts them into length values in a consistent way. Then we can employ a modified form of the Pythagorean theorem—the geometric relationship that relates the hypotenuse and sides of a right triangle—to find the space-time interval.

Technically, the procedure is as follows: Take the spatial distances in all three directions and square them. Next, take the time difference, multiply by the speed of light, and square the result. Finally, subtract that value from the sum of the squares of the spatial distances, yielding the square of the space-time interval.

Notice that the spatial distances are added, but the time difference (multiplied by light speed) is subtracted. The procedure that governs which terms to add and which to subtract is called the signature. In standard Euclidean geometry, of the sort Einstein studied as a child, all distances are additive. Hence, the signature is fully positive. In Minkowski space-time, on the other hand, the temporal "distance" is subtracted, yielding a mixed signature of three "plusses" (for space) and one "minus" (for time).

The mechanism, in general, to determine the space-time interval for any given set of events and region of the universe is called the metric. For Minkowski space-time, the metric is relatively simple: Add the squares of the spatial terms and subtract the square of the temporal term (multiplied by the speed of light). It is known as a "flat" metric—"flat" indicating that the shortest distance between two points is a straight line, not a curve. However, as we'll see, other metrics are decidedly more complex.

The opposite signs of the space and time parts of the Minkowski metric indicate that the space-time interval can be positive, negative, or zero. These have three distinct meanings. In the first case, positive, the spatial terms dominate, and we call the interval "spacelike." A spacelike interval means that causal communication is impossible because there is simply too much spatial separation for a signal to travel in the given interval of time. If, in contrast, the interval is negative, it is called "timelike." In that case, the temporal dominates the spatial, and signals have more than enough time to make the journey. Finally, a third case, known as a "zero" or "lightlike" interval, refers to the exact amount of space crossed by a light signal in a given time. This does not, of course, imply zero separation in three-dimensional space. Rather, it enshrines the special status of light rays as the quickest connections through space-time.

So in our futuristic scenario, if a café on Ganymede orders fresh supplies of coffee beans specially grown on Venus, almost half a billion miles away, there must be a minimum time delay between the order and shipment of roughly three-quarters of an hour for these events to be causally connected. This allows enough time for communication (by radio or other means) to occur. Therefore, if the order is placed at 11:45 a.m. and the shipment goes out at noon, the time delay would be too brief for a signal to have traveled from one place to the other. The interval would be spacelike, and we'd have to chalk up the sequence to pure coincidence. If, on the other hand, the shipment leaves the following day, the time delay would be sufficient for us to conclude that it was in direct response to the order. The interval, in that case, would be comfortably timelike.

Special relativity is a highly successful theory. A young boy's daydreams, polished by years of painstaking calculations, have delivered an extraordinarily accurate description of near-light-speed dynamics.

Its astonishing predictions have proven correct in numerous applications. For example, when placed on high-speed aircraft, ultraprecise cesium clocks lag by the precise amount Einstein predicted. Mammoth accelerators boost elementary particles to near light speeds by faithfully timing their actions to relativity's rhythms. Nuclear reactions generate energetic offspring that—invigorated by time dilation—live longer lives than their slower cousins.

Given such a fantastic achievement, why didn't Einstein stop there? Why did he wrestle with nature's laws for another decade, until he could mold special relativity into a far more mathematically intricate theory, known as general relativity? The reason stems from two critical omissions: acceleration and gravity.

An avid reader of Mach, Einstein knew that special relativity failed to answer Mach's question, "How does it come about that inertial systems are physically distinguished above all other coordinate systems?" That is, what makes constant velocity the favorite type of motion in nature?

Einstein also realized that this question was deeply linked to the mysteries of gravitational attraction. Why, without air resistance, do light feathers and heavy stones drop toward Earth at the same rate? Clearly, he surmised, gravity's pull cannot just depend on the bodies in question but must be seated in space-time itself.

THOUGHTS IN FREE FALL

In similar fashion to his earlier theory, gravity came to Einstein's attention in the form of a thought experiment. He imagined someone falling off the roof of a house while simultaneously dropping an object (say a box of tools). As the unfortunate man descends, he notices that it remains right next to him. Although the dropped object is falling independently, he can reach out and grab it whenever he wants. Except when he hazards to look down, his situation seems to him exactly as if he had remained at rest. That is, of course, until he and the object simultaneously hit the ground with a thud.

These musings led Einstein to posit a fundamental property of nature, called the "equivalence principle," that governs what happens when objects fall freely due to gravity. It states that no physical experiment can distinguish between free-falling motion and the state of rest. For example, astronauts plummeting toward Earth in a windowless spaceship could well imagine that they are floating in deepest space. Unless they fired the ship's braking rockets, they would notice no extra forces that could distinguish the two situations. For all intents and purposes, they would be resting in an inertial frame.

Hurling a mighty rock through Newton's stain-glassed vision of eternity, Einstein's brilliant proposition shattered its unified concept of inertia into myriad parts. No longer could science consider the state of constant linear motion to be a global property. Rather, it would depend on the gravitational field at any point in space. (A field is a point-by-point description of how forces act on objects.) Like a patchwork quilt, the fabric of the universe would henceforth consist of local free-falling frames sewn together. Each segment would represent inertia according to the immediate gravitational conditions. A piece near Earth would describe free motion in terrestrial gravity, for example; a piece near Jupiter would describe Jovian gravity. All that remained, Einstein realized, would be finding the thread to stitch these fragments together. But that would not be easy.

For a number of years, while working in Zurich, Prague, and Berlin, Einstein wrestled with the difficult issue of connecting the coordinate systems of disparate parts of space-time. Realizing that solving this problem would require potent mathematical machinery, he turned to his close friend, mathematician Marcel Grossmann, who guided him through the nuances of higher geometry. Finally, in 1916, Einstein completed and published his general theory.

Like its antecedent theory, general relativity is four-dimensional. However, it is more flexible and far-reaching than special relativity.

It deals with the varying types of motion caused by the attraction of any kind of matter or energy. Representing Einstein's comprehensive theory of gravitation, it describes how materials produce and respond to changes in space-time geometry.

To achieve this more general theory, Einstein had to move from flat Minkowski space-time to what are known as Riemannian manifolds. These are named after Bernhard Riemann, a German mathematician who died in 1866 at the early age of 39. Manifolds are multi-dimensional geometric representations that can twist and turn like flags in the wind. Unlike Minkowski space-time, Riemannian manifolds can bend at any given point, turning straight lines into curved paths.

Mathematicians express this curvature using several related expressions, technically known as "tensors." A tensor is an object that undergoes specific predictable changes whenever a manifold's coordinate system is transformed (rotated, for example). Therefore, like perfect lenses, they provide consistent images no matter which way they are turned. Because of this regularity, tensors offer an ideal way of characterizing geometries.

Einstein's formalism refers to a number of tensors. The most complete measure of curvature is called the Riemann tensor. It can be reduced into another object called the Ricci tensor, named for Italian mathematician Gregorio Ricci-Curbastro, the founder of tensor calculus. Add another term, and this becomes the Einstein tensor. Finally, all these expressions are related to the metric tensor— the generalization of Minkowski's space-time interval to curved manifolds.

Einstein hoped that one of these expressions for curvature could be directly linked to the matter and energy in any particular region of the universe. Indeed, the Einstein tensor does this job quite nicely. Einstein's general relativistic relations equate it to yet another tensor, called the stress-energy tensor, which characterizes the material and energy content of each part of space. In other words, these tensors connect the bending of space-time with the nature of the substances within it. Generally, the more mass a region contains, the greater its warping, like a field of snow trampled deeper by heavier boots.

From the matter comes the curvature. This, in turn, affects the metric. Unlike Minkowski's metric, of simple plusses and minuses, in the general case each distance or time term is multiplied by an independent factor. These factors can vary from place to place and from moment to moment. They respond to the conditions in a particular locale. Thus, in short, the changing distribution of material in a region alters its web of connections between points, leading to new avenues of motion.

We can imagine special relativity, described by Minkowski's metric, as a staid rectangular building, constructed of uniform rows of identical steel girders. Each girder joins the other in perfect perpendicular fashion, maintaining the same shapes, sizes, and relationships forever. If the universe were like this, it would be as homogeneous as a 1960s public housing project. Moreover, it would have no gravity, since all paths through space would be endless straight corridors.

A more general Riemannian manifold, in contrast, has a far more flexible structure, echoing the complexity of the actual universe. Depending on the coefficients set forth in its metric, each of its girders can vary in size from point to point. Over time they can shrink or expand, becoming indefinitely small or unimaginably large. The result is an elastic architecture more akin to the lithe, flowing creations of Spanish designer Antonio Gaudi than to conventional buildings. Indeed, it is an architecture malleable enough to model the evolving dynamics of an intricate cosmos.

Just as tourists weaving through *La Sagrada Familia*, Gaudi's sinuous church in Barcelona, must take more convoluted paths than if they were traipsing down a flat, straight sidewalk, objects in Riemannian space are often forced through circumstance into curved trajectories. This is true for the planets of the solar system, as they follow elliptical paths around the Sun. While, according to Newton's theory, the gravitational pull of the Sun *breaks* the planets' natural inertial states, in Einstein's theory they *are* in their natural states. The mass of the Sun warps space-time, changing, in turn, the motion of the planets. Therefore a "straight line"—or more properly a "geodesic" (most direct path)—in Riemannian space-time may not look straight at all in ordinary space.

Although the real world is one of curved space, for sufficiently small regions of the universe (such as laboratories on Earth), Newtonian and Einsteinian theory barely differ in their predictions. Experiments done on particle accelerators such as CERN (European Organization for Nuclear Research) in Switzerland do not need to take general relativity into account. Because tiny regions of space-time are essentially flat, they can be modeled well by Newtonian mechanics for low speeds and special relativity for near light speeds. Nevertheless, Einstein proposed several key tests of general relativity that could distinguish it from other theories for sufficiently curved regions. These tests involve the most warped part of the solar system: the region closest to the Sun.

Einstein's first prediction concerned the orbital precession of the planet Mercury. It was well known that planetary orbits don't stay in place forever; rather they advance slightly each time, like the minute hand of a clock. The gravitational theories of Newton and Einstein offer somewhat different values for this rate. Therefore, Einstein was pleased when he discovered that his prediction was more accurate.

Despite this success, Einstein realized that his theory required a stronger test to distinguish it from other possible theories of gravity. Determining that the Sun's gravitational well would be sufficient to bend starlight, he hoped that astronomers would find a means to measure this effect. Such efforts were delayed, unfortunately, because of the poor state of international cooperation during the First World War.

In 1919, with the war finally over, British astronomer Arthur

Eddington organized two expeditions to the southern hemisphere to record effects on starlight during a total solar eclipse. At totality the Sun's rays would be completely occluded for several minutes, giving an observer enough time to examine the Sun's warping of the space around it by measuring the bending of light rays from stars near the edge of the Sun's disk on their way to Earth. Eddington himself led one of the teams down to the island of Principe off the coast of Africa. The other group, serving as backup, went to Sobral in Brazil. The backup plan proved most fortunate when Eddington's voyage turned out to be literally a wash. A deluge of rain drenched Eddington and his team members as they tried to make out the stars through the clouds. They did manage to take some photos, but the ones from Brazil were generally much clearer. Merging these results, Eddington calculated the bending. It agreed reasonably well with general relativistic predictions.

A third test of general relativity, called gravitational redshift, involved the light emitted by the Sun itself. Resembling Doppler shifts, the predicted effect pertained to the reddening of light escaping a deep gravitational well. According to Einstein's theory, the strong gravitational field near the surface of the Sun should slow down the rate of clocks there, resulting in the lowering of luminous frequencies. This postulate can be tested by looking at the spectral lines of atoms—which act like tiny timepieces. The same process should occur near the surfaces of other stars—especially compact ones like Sirius B, the white dwarf companion to Sirius A (the Dog Star). Unfortunately, while the Sun and other bright stars are easy to observe, the physics of their churning surfaces is hard to decipher. So this test was less clear-cut than either Mercury's precession or the Sun's light bending. Sirius B's redshift would not be measured until the mid-1920s.

However, by 1919 the weight of the data for the other two tests was clearly in favor of Einstein's theory. Eddington, who at that time was one of the few people in the world who properly understood it, announced that general relativity was right. Headlines around the world proclaimed the death knell of the Newtonian age and heralded the debut of Einsteinian physics.

A TROUBLESOME MARRIAGE

Today, the scientific community considers general relativity the most accurate and elegant description of the workings of gravity. Nevertheless, many theorists take issue with some of its profound limitations. The foremost of these quandaries concerns its lack of any obvious connection to quantum mechanics—the other physical revolution in the early 20th century.

Of the four fundamental forces of nature, three have been well interpreted through quantum principles. Physicists have combined electromagnetism and the weak interaction (the force that precipitates nuclear decay) into a unified quantum field theory, called electroweak theory. Researchers have similarly modeled the strong interaction (the force that binds protons and neutrons in atomic nuclei) through a theory known as quantum chromodynamics. Yet gravity, the fourth force, remains the odd man out.

Given the vastly different scopes and methodologies that separate quantum mechanics from general relativity, it is no wonder that the search for a full quantum treatment of gravitation has proved elusive. While quantum theory deals with the lilliputian domain of elementary particles, general relativity concerns itself with stellar and galactic behemoths, as well as the vast cosmos itself. Quantum mechanics proclaims, through its famous uncertainly principle, the impossibility of knowing the exact positions and velocities of any object at the same time. This won't go for relativity, which requires such information to render predictions. Moreover, while the quantum world generally relies on a fixed background in space and time, general relativity incorporates space and time into its very dynamics. Thus, while quantum physics conducts its mysterious drama on the space-time stage, Einsteinian gravitational theory is pulling the carpet out from under its feet.

Early attempts to fashion a quantum theory of gravity were further stymied by the presence of mathematical monstrosities, called "infinities," in the basic equations. These anomalies stem from trying to consider tinier and tinier regions, eventually homing in on exact geometric points. By dividing such infinitesimal distances, one is left with indeterminate expressions. For the other forces, physicists have found ways of canceling such problematic items, but not so for gravity. Gravity, considered on its most miniscule scale, is plagued with unavoidable infinite terms that render attempts at calculation meaningless.

A clever way of handling this situation derives from modern string and membrane theories, which posit that point particles do not even exist. Rather, they theorize, the smallest units in nature are vibrating strings and sheets of energy. By excluding mathematical points, these theories abolish the infinities from quantum calculations. For this reason many theorists believe they offer the correct pathway to quantum gravity.

Such theories predict that for minute distances probed at ultra-high energies, gravitational behavior would begin to deviate from standard general relativity. Einstein's equations (relating space-time geometry to its material content) would accrue extra terms, leading to measurably different results. Thus, gravity would have two different faces, its familiar visage seen in the ordinary motions of stars and planets and an exotic countenance discernible only under extreme circumstances.

Where might such a hidden face be found? Perhaps in the fiery first instants of the universe, gravity could scarcely be distinguished from the other natural interactions. Maybe, as physicist John Wheeler once proposed, the early cosmos was a space-time foam—a jumble of free-flowing geometry leaping from one quantum configuration to another. In those turbulent moments, gravity and the other forces could have continuously exchanged properties and identities, energetically exploring myriad characteristics. Within this swirling amorphous billow of inconceivably intricate connections, even the number of spatial dimensions could have varied wildly. Reality, if we could somehow perceive its earliest state, would have been a maddening labyrinth.

Then, as the universe cooled down, each natural interaction might have locked into place. One by one, like crystals slowly assembling on a watery surface, each force would assume its permanent form. As space-time's froth turned more solid, gravitational behavior would settle into its current profile. Finally, like an icy lake in winter, the ripples and eddies of sultrier times would be completely frozen over.

Obviously, we cannot travel back in time and experience the nascent cosmic conditions ourselves. But perhaps sifting through current astronomical data could somehow reveal aspects of this embryonic development—much like a doctor surmising from a child's health what his fetal environment may have been like. Or maybe powerful particle accelerators, such as CERN's Large Hadron Collider scheduled to go on line in 2007, will replicate the high temperatures of the early universe and produce discernable effects.

Alternatively, we could hunt for regions in space where matter is dense enough that conventional general relativity could break down. Astronomers believe such conditions might be present in the massive remnants of collapsed stellar cores—ultracompact objects known as neutron stars and black holes. Within their shrouded interiors, the lexicon of ordinary physics could give way to an unknown language so bizarre as to be barely comprehensible.

IN THE DEPTHS OF A BLACK HOLE

When Einstein developed the equations of relativity, he hoped they would resolve the dilemmas posed by Newtonian physics without generating new problems of their own. Ideally, he envisioned an airtight description of the cosmos without any open ends. A strict determinist in the tradition of philosopher Baruch Spinoza, Einstein expected that a full accounting of nature would prove unambiguous and unique. A deity, the esteemed physicist argued, would have no reason to create an imperfect universe with any aspect subject to chance or interpretation. "God does not play dice," he famously remarked.

Ironically, however, loose threads began unraveling from Einstein's supposedly seamless garment almost as soon as he had fashioned it. One of the first general relativistic solutions, calculated by German astrophysicist Karl Schwarzschild in 1916, possessed a curious open end, called a singularity, that seemed impossible to remove or explain. A singularity is a point or region where certain parameters (such as density or pressure) zoom to infinity, creating a breach in the fabric of space-time. Einstein deplored singularities because they rendered theories mathematically incomplete.

Schwarzschild, officially the director of the Potsdam Astrophysical Observatory but then serving on the Russian front as an artillery expert, developed his solution to describe the relativistic properties of stars. He modeled stars as spheres of particular sizes and masses. Churning these variables through Einstein's equations, he obtained a metric describing the warping of geodesics ("straightest" paths) near such bodies.

Physicists often like to test-drive solutions by exposing them to extreme conditions. In the case of the Schwarzschild metric, this involved imagining what would happen if the star's mass was high but its radius extremely small. Strangely, this changed its character from a simple dent in space-time to a bottomless pit. Beyond a certain point, called the "event horizon," geodesics entering this region would no longer be able to escape. Hence, light rays—traveling along geodesics—could enter but never leave. Today, we call this situation a "black hole"—so dubbed by John Wheeler for its light-trapping properties.

Since the 1960s, when Wheeler introduced the expression,

astronomers have identified a number of black hole candidates. One might wonder how they can detect such coal black objects against the backdrop of darkest space. Like a ghost sitting on a seesaw and lifting a startled child resting on the other end, astronomers have sensed these unseen bodies through the reactions of those around them. Many black hole candidates have been found in binary star systems by noting their actions on visible stars. Black holes are thought to victimize their companions by absorbing their material in a process called "accretion." As such captured matter plunges into the black hole's bottomless gravitational well, it reaches ultrahigh temperatures, causing it to emit highly energetic radiation, mainly in the form of X-rays. Astronomers have recorded such characteristic signals, leading them to conclude that black holes likely exist.

Black holes, according to current thinking, comprise one of three possible end points for stellar evolution. When a star's primary source of energy—its nuclear fuel—becomes exhausted, its central core collapses and its outer envelope expands. The peripheral material exudes into space—either in a gradual dissipation (for lighter stars) or in a catastrophic supernova explosion (for heavier stars). In the former case, the remaining core settles down into a hot, tiny beacon, called a white dwarf. Such will be the fate of the Sun.

A star between 1.4 and 3 times the mass of the Sun, however, suffers a far more turbulent fate. Its core implodes so suddenly and energetically that the very atoms inside it are completely pulverized. Throughout the collapsing body, positive protons and negative electrons fuse into neutrons. This happens simultaneously with the supernova explosion of the outer shell—similar to the pulling back of the undertow when ocean waves are building up. The core—an ultradense amalgamation of neutrons known as a neutron star—remains as a relic of the catastrophe.

Finally, if a star is more than three times heavier than the Sun, its violent transformation is even more powerful. Not just the core's atoms but also its elementary particle constituents are utterly destroyed. Nothing remains of matter as we know it. What's left is an infinitely dense singularity cupped by a deep, light-trapping gravitational well—in other words, a black hole.

Despite promising candidates and sensible formation theories, scientists can only speculate about a black hole's shrouded interior. The region between a black hole's central singularity and its event horizon constitutes perhaps the most enigmatic frontier in modern astronomy. General relativity advises us that a series of extraordinary, but ultimately deadly, events would transpire for any brave or foolish soul who dares to venture within its ghastly domain.

A black hole would be a most insidious snare to anybody entering it, for sure, as it would give little warning of the perils in store. At first, astronauts on such a doomed spacecraft would feel nary a jangling of their silverware as they approached the dark, frozen object. Looking at their watches, they'd notice nothing of particular interest, little knowing that their timelines were rapidly diverging from those on Earth. The reason for such a discrepancy is that the time axis tilts in regions of gravitational distortion. This variation in the direction of time's axis resembles the twisting of the quills of a porcupine, pointing in different ways on various parts of its curved body. The tilting of time's axis near the black hole contrasts with its "upright" direction in relatively flat regions far away from it, leading to a comparative dilation of time—similar to special relativistic effects but due to gravity rather than high speeds. So as time passed normally for the unfortunate crew, those following their travails from a safe distance (we imagine here a remarkably powerful telescope observing the ship) would be horrified to see them moving more and more slowly. Like characters from a George Romero flick, they would seem like languid automata inching their way across the deck of an increasingly dormant vessel.

Eventually, as far as the outside world is concerned, the ship and its occupants would grind to a virtual halt at the brink of the black hole's event horizon. Their clocks would be moving so slowly, relative to Earth's, that for all intents and purposes they'd be statues. Not so, however, from the astronauts' perspective. Time would continue for them unabated as they sailed through the invisible barrier. From that point on, there would be no turning back. To escape, they'd have to reverse course at a rate faster than light—an impossibility.

What would transpire next for the fated passengers depends on the size and nature of their captor. Physicists have generalized Schwarzschild's simple model to encompass more elaborate possibilities. Additional black hole solutions have been found, representing spinning and electrically charged varieties. The complete description of a black hole state, named the Kerr-Newman solution for theorists Roy Kerr and Ted Newman, delimits all possible masses, sizes, rotational rates, and charges.

A curious expression coined by Wheeler, "black holes have no hair," designates physicists' opinion that these are the *only* parameters that have meaning for such bodies. Everything else notable about them (such as the specifics of their origins) would be shorn off by relativity's meticulous barber. Cruelly, this would also be true for anything or anyone that happened to be ingested. There'd be no mark or tattoo on a black hole's bald pate that would indicate its contents.

Still, given their wide range of possible masses and rates of spin around their axes (as well as whether they are electrically neutral or charged), not all black holes are the same. Candidates have been detected with an enormous variety of sizes—ranging from large stars to the central dynamos of galaxies. Primordial black holes, born from density fluctuations in the early universe, could have been as light as 1/100,000 gram.

More massive objects tend to form bigger black holes. For example, a black hole three times the mass of the Sun would have a Schwarzschild radius (distance from the singularity to the event horizon) of approximately 5 miles, the size of a small city. In the center of the Milky Way, by comparison, there may be a black hole estimated to be more than 3 million times as massive as the Sun. Its Schwarzschild radius is thought to stretch out almost 5 million miles—or 11 times the radius of the Sun.

Variations in size would have major impact for our trapped astronauts. A small black hole would almost immediately crush them—offering them not even a moment's respite to contemplate their fate. If, on the other hand, they were "lucky" enough to fall into a large black hole, they would have ample time to soak in their surroundings—a flood of lethal radiation—while taking a gut-wrenching plunge to its center. As they sank into the abyss, tidal forces would stretch them out along their path of motion while squeezing them like a tube of toothpaste in the other directions. In either case, quick or slow, the ultimate result would be a complete pulverization of every molecule in the astronauts' bodies.

One is reminded of the scene in the film *Arsenic and Old Lace*, when mad Dr. Einstein (played by Peter Lorre) decries his cohort's decision to apply slow torture instead of quick murder to the cap-tured protagonist (played by Cary Grant). The trembling plastic surgeon begs his coconspirator to just get the killing over with. "Not the Melbourne method!" he pleads to no avail. "Two hours!" Never-theless, the choice of a two-hour technique offers the leading character precious time to be rescued.

Given sufficient time, could astronauts find a way to escape a black hole's crushing singularity? That would depend on whether a highly theoretical conjecture about such objects turns out to be true.

PORTALS TO THE UNKNOWN

On the face of it, a black hole represents a one-way journey to a crushing death. But that's just the classical picture. According to quantum notions, captured material does slowly leak out—in a trickle of energy, called Hawking radiation, that exudes from the event horizon over the course of trillions of years. Whether or not such leaked energy could convey information about the original objects is still controversial. For decades, Cambridge physicist Stephen Hawking, the developer of the theory, argued that it does not. During a recent talk at a scientific conference, however, he indicated that he has changed his mind. Bits of information, he now believes, could be released in the trickle. Nevertheless, because it would be painstakingly slow and would not constitute information on the actual original bodies that were sucked into the black hole, this method of "escape" would hardly be comforting to trapped astronauts about to be pureed.

Of greater possible interest is the notion of "tunneling" intact through the black hole to another part of space-time through a type of interconnection called a "wormhole." This hypothetical link between disparate segments of the universe appears when the Schwarzschild metric and other black hole solutions are plotted on special charts, called Kruskal diagrams, that convey their causal structures. These diagrams suggest that a black hole's seemingly bottomless funnel might not be bottomless at all. Rather, it could be connected via a space-time "throat" to a second funnel. Just as matter would vanish without trace into the first funnel, it would materialize without sign of origin from the second. Theorists have deemed the all-emitting second funnel a "white hole," to contrast it with its all-absorbing polar opposite.

Given their hypothetical nature, the greatest use of wormholes so far has been as a plot device in science fiction stories. Speculative writers had long sought a rapid transit system for conveying terrestrials and aliens from one sector of the cosmos to another—a kind of "subway to the stars." With conventional space travel so slow, wormhole connections appeared a far superior solution. Readers or filmgoers loved to suspend disbelief and take wild rides through interspatial tunnels to worlds unknown.

Ironically, one such science fiction drama stimulated bonafide scientific discussion about wormholes. In the early 1980s, astronomer Carl Sagan was preparing to write *Contact*, a novel envisioning the first human-alien encounter. Realizing that such a rendezvous would require quite a hop across space, he contemplated ways of doing so in a reasonable amount of time.

"That was my problem," recalled Sagan. "To get [the female protagonist] to a great distance away from Earth in the Milky Way galaxy, to meet the extraterrestrials, come back and do all that within the lifetime of the people she has left behind." Sagan knew that black hole tunnels had been discussed as possible gateways but didn't think they'd be safe or feasible. He decided to ask his friend, Caltech astrophysicist Kip Thorne, for advice.

"In the early 1980s there was a common misconception that you might be able to travel from one place to another in the Galaxy, without covering the intervening distance, by plunging into a black hole," continued Sagan. "But there was something about the whole idea that made me nervous. It was for that reason that I contacted Kip Thorne."

When Sagan called him, Thorne confirmed that, although black holes theoretically offered the possibility of interspatial connections, space travelers would be strongly advised not to attempt them. Like a tunnel through an active, lava-filled volcano, such a shortcut would almost certainly prove lethal. Stretched out like taffy, bombarded like in a microwave oven, accelerated like on the most evil thrill ride imaginable, no sane person would wish to buy such a ticket—even if they could somehow get a chance to meet E.T. They might as well go to Universal Studios—which, unlike a black hole, is safety inspected.

Thorne wondered, though, if a more user-friendly wormhole could be developed. Along with graduate student Michael Morris, he examined how a black hole could be modified to eliminate its deadly features while preserving its potential to connect with other regions of space. Tinkering with various solutions to Einstein's equations of general relativity, they managed to fashion a streamlined "traversable wormhole" that would permit safe passage between one region of the universe and another. They sent the results to Sagan, who incorporated them into his novel. In 1987 they published these findings in the *American Journal of Physics*—hoisting the issue into the mainstream of theoretical discussion. Shortly thereafter, New Zealand physicist Matt Visser (then at Washington University in St. Louis) developed an alternative set of navigable wormhole models—proving that there were many ways to carve stable tunnels through space.

Before submitting any engineering bids just yet, any prospective wormhole entrepreneur should stop and consider the enormity of such an undertaking. Constructing a wormhole would require the technological know-how of a civilization far more advanced than ours. Gargantuan amounts of material—many times the mass of the Sun—would need to be compressed and molded into ultracompact configurations. Such a colossal enterprise—assuming it's even possible—could easily be many centuries away.

Furthermore, in addition to the immense technical challenges, the traversable wormhole models all share one major catch: Stabilizing the wormhole's throat would require a special kind of substance, dubbed "exotic matter," with repulsive rather than attractive properties. Like scaffolding holding up a coal mine's ceiling, exotic matter would keep the tunnel from caving in—allowing astronauts to pass through without being crushed. So why couldn't scientists simply find or create such material? The tricky point is that, unlike any of the familiar substances around us, exotic matter would, under certain conditions, be observed as having negative mass. A ripe hanging apple tossed by the wind will eventually fall to the ground. But a negative mass apple would rise to the clouds instead. In other words, it would weigh less than zero.

LIGHTER THAN WEIGHTLESS

How could something weigh less than zero pounds? Could such strange fruit exist? Are there watermelons somewhere in the universe that would levitate from grocer's scales? Are there negative mass boxes of chocolate-covered cherries that would actually remove poundage with each serving? After enough bites, could we float like Mary Poppins? Despite numerous experiments, scientists have yet to detect particles with negative mass. Even positrons, the oppositely charged antimatter counterparts to electrons, have positive mass. Experiments at the Stanford Linear Accelerator have confirmed that positrons indeed fall down, not up.

Curiously, the laws of gravitational physics—whether expressed in Newtonian or general relativistic form—don't explicitly rule out the existence of negative mass. Therefore, following the dictum (attributed to physicist Murray Gell-Mann) "Whatever isn't forbidden is compulsory," surely it must lie somewhere. British astronomer Hermann Bondi once speculated that every positive mass particle could possess a negative mass companion, just as magnetic north poles must waltz with south poles. Then where are these sub-weightless creatures hiding? Could they be huddled in some remote corner of space—banished to the universe's Siberia through sheer gravitational repulsion? Or could they reside closer to Earth, albeit in some dim attic of possibilities we have yet to explore?

As it turns out, you wouldn't need all that much exotic matter to prop open a wormhole. In 2003, Visser and two colleagues calculated that the spatial vacuum—the fuzzy realm of fluctuating quantum fields where uncertainty reigns supreme—could well provide such material. As modern quantum theory has shown, no vacuum is truly empty. The Heisenberg uncertainty principle, a key element of quantum physics, permits particles to materialize from sheer nothingness, as long as they remain only for brief intervals. Conceivably, through this process, tiny amounts of negative mass could randomly emerge from the void. Normally, these bits of flotsam and jetsam would return to the great emptiness, but perhaps they could somehow be captured first. If just a smattering could be netted, Visser's team showed that it would suffice to keep a wormhole's throat open. Like jalapeño sauce, just a few potent drops would be more than enough.

Another potential place to fish for negative mass would be in the deep space-time troughs of neutron stars and black holes. Near the packed centers of collapsed stars, where gravity wears titan's boots, the conventional laws of physics might be well-enough trampled to permit small quantities of exotic matter to leak out. To detect such elusive material, we'd need to drop an enormous test object (like a planet) into a stellar relic and measure precisely what happens. As the test body plunged into the well, theoretically the negative mass would reveal its presence with a characteristic echo.

Some of the theoretical models permitting negative mass involve extending Einstein's equations by an additional dimension, thus augmenting the four dimensions of space-time by one more. Dating back to an early 20th-century proposal by German mathematician Theodor Kaluza, extra dimensions have become a popular avenue for enlarging the scope of general relativity and encompassing electromagnetism and the other natural forces in a unified theory.

Traditional higher-dimensional theories, including Kaluza's, are usually designed to forbid any influence of extra dimensions on the known laws of physics. For example, in a model proposed soon after by Swedish physicist Oskar Klein, the fifth dimension is curled up so tiny that it could never be observed. Like the minute stitches on a finely woven sheet, space-time would feel perfectly smooth to the touch—without indication of something extra.

However, more recent unified theories (such as particular versions of what is called M-theory) involve large extra dimensions— new directions that aren't twisted up into miniscule knots. Rather, the additional dimensions lie along pathways that cannot be accessed by conventional matter but can still make their presence known. The theory allows for physical tests by indirect means.

Intriguingly, the possibility of negative mass could be used as one way of detecting extra dimensions. Particular solutions of general relativity, extended by an extra dimension, display curious sensitivity to the sign (plus or minus) of a particle's mass. They offer stark predictions for what would happen to objects under extreme gravitational circumstances—for example, near the event horizon of a black hole. The existence of negative mass would produce characteristic behavior that astronomers might be able to measure.

The same contemporary higher-dimensional theories offer another startling prediction. Not only do they distinguish between negative and positive mass, they also differentiate between two different uses of the term "mass" itself. They yield distinct values for "inertial mass," a body's resistance to forced changes in motion, and "gravitational mass," which causes a gravitational field. Both originate in Newtonian physics, albeit in separate equations. The former enters into Newton's second law of motion—force equals mass times acceleration—and the latter into his law of universal gravitation. Newtonian mechanics, though, treats these formulations of mass as if they always have the same values. It uses just one variable for these two concepts. Einsteinian general relativity goes even further. The equivalence principle on which it is based mandates that inertial mass and gravitational mass are identical. But what if they were slightly different?

Imagine that you often hear about someone named Moe. First, your next-door neighbor tells you that Moe trimmed some of the trees on your block. Then you learn from the couple around the corner that Moe plowed your street after a snowstorm. You might well conclude that the same handyman did all this work.

Then you find out that the guy who trimmed the trees is tall and has long blond hair and a stick-thin build. After also hearing that the snow plower is short and has curly dark hair and a paunch, you would realize there are two different Moes. Could there really be two different types of mass that do two different jobs?

General relativity describes gravity as a geometric effect in four dimensions. Particles move through space independently of their mass. Wispy neutrinos and bulky upsilon particles, acted on only by gravity, must travel along identical paths because the inertial mass and gravitational mass, being precisely equal, do not enter into the equations of motion.

However, once a fifth, uncurled extra dimension supplements space-time's ordinary four, standard general relativity undergoes a profound transformation. For solutions of Einstein's equations in five dimensions, an extra force rears its head. This new force depends on the motion of ordinary space-time along the fifth dimension. Moreover, it accelerates particles as a function of their mass, clearly violating the equivalence principle. It would cause, for example, two asteroids plunging toward the Moon—one large, the other tiny—to fall at slightly different rates.

Questions about the absolute validity of the equivalence principle and other issues concerning the nature of gravitation have stimulated a number of experiments designed to test the fundamental assumptions underlying general relativity. Given that Einstein's marvelous theory is now a proud nonagenarian, perhaps she could use some checkups to gauge her health. Will she continue to be the beloved grand dame of modern physics, or will one of her offspring assume her exalted position? Bets are on her continued survival, but it will be interesting to see what the prognoses reveal.

BALANCING ACT

General relativity is, by its very nature, harder to test directly than other physical theories. Unlike laboratory-based disciplines such as biophysics or materials science, its focus is far more remote and less tangible. We can't simply place the fabric of the universe under a microscope to see if it obeys certain geometric relationships. Unlike, say, an unknown metal, we can't pound space-time with a hammer, press it with a die, or stretch it out on a roller to ascertain its tensile properties. Nevertheless, researchers have devised subtler methods of putting it through the wringer.

The results from the 1910s—the perihelion advancement of Mercury and the behavior of starlight near the Sun—were important early gauges of relativity's overall viability—akin to making sure a patient has a reasonable heart rate and blood pressure. Another critical test, the existence of gravitational redshifts, similarly checked out fine. But by the 1950s many researchers expressed dismay that no more tests were available. For example, at a 1957 conference, physicist Bryce DeWitt threw a piece of chalk up in the air, caught it, and then remarked (slightly exaggerating): "We know almost nothing about gravitation. There is only one experiment which we do over and over again, and that is what I have just done." Fortunately, a bevy of new probes now offer Einstein's body of work an even more extensive physical examination. In assorted experiments, precise equipment has been scanning it from head to toe, seeking signs of even the slightest flaw.

Providing the very legs on which relativity stands, the equivalence principle must remain solid enough to support the theory. Accurate measurements of the equality of inertial and gravitational mass offer vitally important data. If they were to indicate even the slightest discrepancy, the implications would be monumental. Modifying Einstein's theory would become a necessity, not just speculation.

One device for testing equivalence, called a torsion balance, dates further back than general relativity itself yet continues to be updated and refined. Torsion means twisting or turning. Through a balance device, such actions can reveal how forces affect materials. At the turn of the 20th century, Baron Roland von Eötvös of Hungary used such a sensitive instrument—a weight hanging from a rotating rod delicately balanced on a pivot—to measure minute differences in the accelerations of various substances. He devised it to record any subtle effects produced by small discrepancies between inertial mass and gravitational mass. Thanks to its meticulous design, the equipment was precise enough to rule out such a difference down to one part per hundred million.

Eötvös's measurements stood as the benchmark for decades, offering a firm basis for Einstein's assumptions. Then in the early 1960s astronomer Robert H. (Bob) Dicke of Princeton, along with colleagues G. Roll and R. Krotkov, suggested a clever way of substantially improving on Eötvös's method. Realizing that the Sun exerts a periodic pull on terrestrial objects—due to Earth's 24-hour rotation—they measured the accelerations of various objects with respect to the Sun rather than Earth. The device they used was a triangular array of weights: two made of aluminum and one made of gold. An electrical system served to keep the set balanced. If any of the weights felt an extra tug and the device started to tilt ever so slightly, an electrical signal would immediately rectify it. The amount of this signal was carefully recorded.

Now suppose the equivalence principle was false and acceleration depended on mass. Then aluminum would react slightly differently than gold to the Sun's pull. As Earth turned around on its axis, the torsion balance would try to tilt slightly in different directions. The electrical system would thereby need to exert a periodic correction—with an unmistakable 24-hour cycle. Dicke and his co-workers found no such cycle. Within a difference of one part in 100 billion, they confirmed that aluminum and gold accelerate at the same rate under gravity.

Refining Eötvös's concept even further, in the 1990s a group of researchers led by Eric Adelberger of the University of Washington constructed several torsion balances with even greater sensitivity. They designed each balance to test particular features of gravity on a variety of scales. Rather than just looking at the effects of Earth and the Sun, they fashioned their instruments to measure gravitational influences as close as the fly's wings and as far away as the center of the Milky Way. To honor both Eötvös and their university, they named their collaboration the Eöt-Wash group—pointing out that "vös" in Hungarian is pronounced somewhat like "Wash" in English.

Since beginning its experiments, the Eöt-Wash group has delivered an impressive array of data indicating that light and heavy objects accelerate exactly the same way—with a maximum discrepancy of approximately one out of 10 trillion. Gravity, the team has found so far, behaves in an identical fashion, whether it is twirling stars around a galactic core or lowering a speck of dust toward the ground. With these successes in hand, the team is pushing its equipment to its absolute limit, hoping to map out every facet of gravity's terrain.

Today, not all tests of the equivalence principle involve nimble balances twisting and turning in labs. Some of the newer experiments have forsaken Chubby Checker moves for Obi-Wan Kenobi maneuvers. With lasers and space probes now used to make ultraprecise measurements, tests of general relativity have entered the Star Wars age.

The Proof in Space's Pudding

In the 1960s and 1970s, space agencies such as NASA (National Aeronautics and Space Administration) captivated the public through unprecedented manned missions, like the Apollo Moon landings. These days such centers have broadened their scope to include a wealth of scientific satellites and other instrumentation designed to investigate the nature of space itself. The Hubble Space Telescope, the most famous of these instruments, has been joined by numerous other devices probing the deep structure of the cosmos.

Witness a new APOLLO (Apache Point Observatory Lunar Laser-ranging Operation) mission, one that sends laser beams instead of people to the Moon. It makes use of five retro-reflectors—banks of special prisms left behind by the astronauts on the lunar surface. These mirrored surfaces reflect incident light back to Earth, enabling precise measurements of the distance to the Moon. By shining a laser from Earth onto one of these and timing how long it takes for the beam to return, scientists have been able to pinpoint the Earth-Moon distance within a fraction of an inch. Led by astrophysicist Tom Murphy of the University of California at San Diego and including Adelberger as one of the team members, researchers hope to use this method to check for subtle differences between the motions of the Earth and Moon in the Sun's gravitational field. If such discrepancies are found, they could point to minuscule violations of the equivalence principle.

To test the actions of gravity on varying masses, we might wonder why scientists don't just drop two objects and see if they land simultaneously—as, legend would have it, Galileo did from the Leaning Tower of Pisa. The free fall would need to take place in a total vacuum to prevent air resistance from skewing the results, so the project planned by the ESA (European Space Agency), called MICROSCOPE (*MICROSatellite à traînée Compensée pour l'Observation du Principe d'Equivalence*), is designed to do just that. Targeted for launch in 2008, it will reconstitute the leaning tower as a floating satellite, orbiting almost 700 miles above Earth. This vehicle will shield two cylinders, made of platinum and titanium, which will be released simultaneously and allowed to move freely inside. Because both masses will be subject to the same gravitational field, namely Earth's, the equivalence principle predicts that they should follow identical orbits. Each time they deviate from their uniform paths an electrical field will steer them back into place. Therefore, by measuring the electrical signals required to keep both objects moving along the same trajectory, researchers will gain precise information about whether or not the equivalence principle is violated.

In case MICROSCOPE doesn't constitute proof enough of Einstein's conjecture, yet another mission is planned after 2011. Known as STEP (Satellite Test of the Equivalence Principle), it is a joint project of NASA and the ESA. Housed within an Earth-orbiting satellite, hollow test cylinders of various masses will be stacked inside each other like Russian dolls and then placed in a cryogenic (ultracold) vacuum flask. Superconducting shielding will protect the apparatus from external disturbances. (Superconductivity is a low-temperature quantum effect that allows certain materials to maintain electrical currents and magnetic fields indefinitely. It offers a buffer against external electromagnetic influences.) Highly sensitive SQUIDs (Superconducting Quantum Interference Devices) will measure the concentric cylinders' relative motions as the satellite circles through Earth's gravitational field. They'll be able to detect motions as fine as 50-quadrillionths of an inch. Like an overzealous traffic cop, they'll record even the slightest inkling of a violation.

The equivalence principle is not the only aspect of gravitational physics being tested in space. An orbiting satellite called GP-B (Gravity Probe B) is currently engaged in a far-reaching study of two general relativistic predictions: frame dragging and the geodetic effect. These properties, specific to Einstein's theory, are quite subtle and have never before been tested.

Frame dragging involves the twisting of space-time due to the rotation of massive objects. Emanating from each body in the cosmos, like the streamers from a maypole, are manifold geodesics. These strands correspond to the shortest paths through that region of the universe—namely the routes traveled by light rays. When a body twirls around in its clockwork dance, it swings its streamers with it. Objects clinging to these streamers, like May Day revelers, must similarly whirl around, changing directions as they spin.

Although relativistic frame dragging was first postulated by Austrian physicists Joseph Lense and Hans Thirring in 1918, it wasn't until 1959 that Leonard Shiff of Stanford proposed a direct way of testing it. Shiff calculated that a spinning gyroscope orbiting 400 miles above Earth would change its tilt by a fraction of a milliarcsecond (an extremely tiny angle, roughly 300-billionths of a degree) each time it orbits. Though minute, this precession could potentially be detected; it is the main impetus for GP-B.

A second source of tilting, the geodetic effect, arises from the denting of space-time by massive bodies. Dutch scientist Willem de Sitter discovered this property in 1916. When a car drives over bumps in the road, it may swing from side to side. Similarly, if a spinning gyroscope travels through warped space—near a planet, for instance—its axis of rotation tends to lean in various directions. This effect, approximately 6,600 milliarcseconds per year, is also minuscule but decidedly more pronounced than frame dragging.

Both effects are so tiny that we might be tempted to ignore them, or dispute whether they are worth spending our hard-earned tax dollars on testing. However, effects that are small in our solar system can have profound implications for the wider cosmos. For example, Einstein's theory accounts for a tiny change in the orbit of the planet Mercury of 43 arcseconds per *century*. That minuscule result helped confirm the space-bending properties of general relativity—leading, for example, to predictions about massive black holes in the centers of galaxies. Similarly, precise measurements of the frame-dragging and geodetic effects would undoubtedly produce a wealth of new cosmological conjectures.

The GP-B apparatus is specially designed to accomplish this task. Inside an Earth-orbiting satellite is a dewar of superfluid helium, maintaining a temperature of 1.8 degrees above absolute zero. The dewar, in turn, houses a cigar-shaped quartz chamber. Within the chamber are four spinning spherical gyroscopes suspended in an electric field and encased by superconducting lead foil. The extreme cold, electrical levitation, and lead foil each cushion the gyroscopes from stray disturbances. Cold minimizes the random jostling of molecules; levitation minimizes the rocking due to the motion of the vessel; and the foil dampens the influence of Earth's magnetic field. All this ensures that the gyroscopes are steered almost exclusively by gravitational effects.

As in the case of the STEP project, superconductivity plays a second, even more vital role. The gyroscopes are encircled with superconducting loops hooked up to SQUID devices. As they turn ever so slightly, the SQUIDs are sensitive enough to record minute magnetic changes resulting from these reorientations. These devices provide the jeweler's tools needed to examine the fine facets of relativity.

The gyroscopes themselves do not look like the archetypal toy spinning top (though mechanically they act in a similar fashion). Rather, each is a glassy sphere about the size of a Ping-Pong ball, machined to amazing smoothness. The surface of the ball does not depart from that of a perfect sphere by more than a millionth of an inch. To put this into perspective: If Earth were as spherical, its highest mountains and deepest oceans would represent deviations of only about 8 feet!

The GP-B satellite follows a polar orbit 400 miles above Earth. To provide a steady reference point, the orbital plane lines up with a star named HR8703, in the constellation Pegasus. This guide star offers an absolute background against which astronomers can take their measurements. As you can see, nothing about the mission has been left to chance.

The principal investigator of the GP-B experiment, who also happens to be the chief organizer and developer of the STEP project, is Stanford University physicist C. W. Francis Everitt. Born in 1934 in Sevenoaks, a town in the rolling Kentish countryside of south-eastern England, Everitt first learned about general relativity at an unusually young age. One day when he was 12 years old and was sitting at the family dinner table, his father fascinated him with compelling accounts of gravity's actions in the universe. "My father," relates Everitt, "who was an engineer/patent attorney with wide intellectual interests, talked to us about Einstein's and Eddington's popular books on the meaning of relativity. He contrasted these with quantum mechanics which, like Einstein, he found not entirely tasteful."

Everitt did not pursue general relativity as his specialty, however, until 1961 when he assumed a position at the University of Pennsylvania. In that scholarly setting, Stanford physicist Bill Fairbank, a pioneering researcher in gravitational and space science, gave a series of talks describing a number of "far out" experiments. "I found them and him fascinating," recalls Everitt. "Since GP-B was the 'farthest out' of the lot, I volunteered to join his group to work on it. At a deeper level I was also much influenced by a remark [Nobel Prize–winning physicist Patrick] Blackett made to me in London: 'If you can't think of what physics to do next, invent some new technology; it'll always lead to new physics'."

Everitt has remarkable perseverance, given the decades taken for his major projects to reach fruition. From the time he began working on GP-B until the instant it blasted off into space, nine U.S. presidents took their oaths of office, musical tastes ran from Doo Wop to Hip Hop, and the world's population more than doubled. Yet he persisted in his endeavors until he could set his creations free in space.

Convincing NASA to construct GP-B in times of tightening budgets required the skills of an expert salesman. Costing hundreds of millions of dollars, it was the most expensive and technologically ambitious science spacecraft ever commissioned by NASA, and its development became the subject of acrimonious debate in the science community. By and large, theoretical physicists wanted it, while astronomers thought it was unnecessary. Like many things in California, it came to represent a focal point of discontent between those in the north and those in the south. Stanford researchers, from the San Francisco area, wanted it built; while many Caltech researchers, from the Los Angeles area, wanted it scrubbed. An exception in the latter camp was Kip Thorne, who consistently supported the mission and was present at the launch. The experiment came perilously close to being closed down several times by NASA, whose critical visits to the Stanford campus were likened by one senior figure as akin to interrogations by the Spanish Inquisition.

Even the probe's launch, from Vandenberg Air Force Base in south-central California, proved a nail-biting test of patience. From December 6, 2003, to April 17, 2004, the mission was held up because of a revamping of its electronics. Then, a broken cord on the launch tower caused another delay. Finally, on April 19, things seemed ready to roll. The mission's organizers bused several hundred people to the site—including various scientists and journalists keen to catch a glimpse of the historic event. Everything went well until the four-minute mark, when the launch was suddenly aborted due to unfavorable weather conditions. The wind profile at the time was not ideal, and nobody wanted to take any chances. Everybody then went back to their hotels, with some people drowning their sorrows at the local taverns. Everitt, however, did not seem downhearted.

Sure enough, the next day when the launch experts tried again, fate was much kinder. First, the epochal words: "Five, four, three, two, one. . . ." A tense pause and then: "We have liftoff for Gravity-Probe B to test Einstein's theory of relativity in space!"

A bright, unnatural light burst over the semiarid landscape of central California. The ball of radiation shot rapidly into the blue morning sky. Perceptibly later, a swath of ragged noise like an avalanche swept over the assembled onlookers. They appeared not to notice. Their eyes were fixed on the actinic light, now halfway up the sky, which marked the location of the spacecraft. It moved surprisingly quickly, heading out over the Pacific atop its Delta II rocket, accelerating to its final speed of about 17,000 miles per hour. In haste, cameras began to click, and a spontaneous cheer followed the probe upward. Heads craned back, hands sheltered narrowed eyes, and after little more than a minute, the sky was empty again, save for a few startled seagulls.

"We did it!" exclaimed one of the onlookers, Dimitri Kalligas. His tone mixed triumph and relief. Having worked on the mission at Stanford for several years in the late 1980s and early 1990s, he relished his dreams finally coming alive. He had traveled from his native Greece, together with his wife and two children to savor the moment. His eyes remained transfixed on the rapidly disappearing probe, even as his kids started pulling him impatiently toward the waiting shuttle bus. With a heartfelt gesture, he did the sign of the cross and turned away; the spacecraft was in the heavens where it belonged. Perfectly aligned with its guide star, it has been in orbit ever since.

CATCHING WAVES

Another critical test of general relativity doesn't involve space probes; it is taking place right here on Earth. The LIGO (Laser Interferometer Gravitational-Wave Observatory) project is attempting to detect gravitational waves, the elusive ripples in the space-time fabric first predicted by Einstein in 1916. A joint project of scientists from Caltech and MIT, the observatory's detectors began operation in 2001 and have scanned for signals ever since.

Researchers believe cosmic catastrophes, such as supernova explosions or collisions between black holes, generate volleys of gravitational waves. These shock waves are thought to fan out in all directions from such disturbances, rattling any massive objects lying in their paths—in the same way that shops rumble when an elevated train passes overhead. Although they've yet to be found directly, astronomers Joseph Taylor and Russell Hulse have used binary pulsars (pairs of rapidly spinning neutron stars) to show that they are likely to exist. For this work they received the 1993 Nobel Prize for Physics.

The LIGO project was proposed by physicist Rainer (Rai) Weiss of MIT, along with Kip Thorne, Ronald Drever, Rochus Vogt, and other researchers at Caltech. Born in Berlin in 1932 to a politically active family, Weiss emigrated with them at a young age to the United States to escape the terrors of the Nazi regime. Like Everitt, he was not originally trained in general relativity but rather in another branch of physics. Weiss received his Ph.D. at MIT, in the field of atomic physics under the supervision of Jerrold Zacharias.

Zacharias had dedicated himself to building high-precision timepieces based on the predictable rhythms of atoms, an extraordinarily important endeavor with broad implications for a variety of scientific fields. As Weiss related, even Einstein in his final years, while engrossed in the search for a unified field theory, expressed interest in the MIT project to develop such clocks. If such devices could be perfected, one of their possible applications would be precise measurement of the effects of gravitation on time. This would help provide further confirmation of general relativity. Zacharias proudly introduced his project to Weiss.

"Jerrold said to me," recalled Weiss, "that he had made himself a clock called the 'fountain clock,' which was a brand new idea involving tossing atoms high into the air and timing them. The idea was to get a long observation time on the atom. He kept telling me that if we could get the clock running, I would travel to the Jungfraujoch, a scientific observatory high in the Swiss Alps. He would be with his clock in the valley and we would measure the Einstein redshift. That's what set the bee in my bonnet about relativity. But the clock didn't work; it was a total failure."

Nevertheless, Weiss's interest in experimental tests of general relativity only grew. Obtaining a postdoctoral fellowship with Bob Dicke, he learned about attempts to measure gravitational radiation. "With Dicke I did something wacky," continued Weiss. "I worked on a gravitometer to measure scalar waves [a hypothesized mode of radiation] hitting the Earth."

Dicke, a master at cutting through thorny mechanical dilemmas, also instilled in Weiss the value of solid experimental design. Returning to MIT as a professor, Weiss embraced the teachings of his mentors and became one of the world's leading experts in high-precision measurements of gravity.

The capstone of Weiss's career is LIGO. Weiss developed the notion of using a special technique called laser interferometry to track minute movements of matter due to gravitational waves. Interferometry involves focused beams of light with well-defined frequencies (that is, laser beams) traveling along separate paths and then coming together again. The pattern created when the beams reunite provides precise information about the difference in path lengths.

Imagine the laser beam to be a troop of soldiers, marching down a road in perfect lockstep. At one point the band needs to cross a river, traversing two parallel bridges that at first glance appear to be identical. They split into two groups, continuing to march all the while. When they reunite, they realize that half of them are now marching out of cadence with the others. A member of the corps of engineers measures the bridges, and sure enough, one is 10 inches longer than the other. The extra length created the asynchrony. The same thing happens with light if it is forced to take several different trajectories. The results are characteristic interference patterns—bright and dark fringes that indicate where the beams are in or out of phase. The spacing of these fringes pertains to discrepancies between the routes.

Weiss and his collaborators realized that such hairbreadth measures would be needed if science had any chance of sensing the ghostly touch of gravitational pulses. Imagine two black holes colliding thousands of light-years from Earth. The resulting catastrophe would send shock waves through the fabric of space, with these rumbles eventually reaching Earth. Nevertheless, even the signal from such a cataclysmic event would offer only a feather touch on earthly objects. The end points of a yard-long iron rod, if it were completely free to move, would be displaced trillions of times less than the diameter of a speck of dust. Thus, the Caltech and MIT researchers planned LIGO to be miles long (for maximum effect) and calibrated as finely as state-of-the-art technology permitted.

The LIGO detectors, located in the states of Louisiana and Washington, are uniquely designed to record the murmurs of passing gravitational waves. Having two widely separated instruments helps rule out the effects of local terrestrial vibrations, such as miniearthquakes or other rumblings, that could masquerade as true signals. A team of planners selected each location to be as far away from urban noise as possible. No one would like a symphony of jackhammers, a band of tractor trailers, and an ensemble of landing jets to serenade the delicate equipment each day—not when it is listening for the subtler melodies of deep space.

Each detector is L shaped, with two perfectly straight vacuum pipes meeting at the corner. Like a colossal bowling alley, each pipe stretches out 2.5 miles long, with target masses on both ends. The idea is that gravitational waves would roll through the tubes, nudge the targets in each arm, and slightly alter their mutual separations. Along one arm, the masses would be pushed slightly closer, while along the other they'd be jostled slightly farther apart. Twin laser beams, meeting at the corner, would record these relative differences through the mechanism of light inference. The characteristic inference patterns would offer a telltale sign of gravitational disturbances (like from eons-old cosmic collisions) faintly touching Earth.

The direct detection of gravitational waves would be a crowning achievement for Einstein's theory. It would well justify all the time and money spent on detectors and probes. As Weiss emphasized, "Observing gravitational waves would yield an enormous amount of information about the phenomenon of strong-field gravity. If we could detect black holes colliding that would be amazing."

Such observations would offer a window into regions in which Einstein's theory differs most greatly from Newton's. General relativity *is* the foundation of modern-day astrophysics and cosmology. We cannot know if our theories of the cosmos are correct unless we can trust Einstein.

MACH REVISITED

While future experiments may indicate a need for its modification, general relativity remains the gold standard. Yet despite its mathematical elegance and predictive success, some physicists are disappointed that it has never fully incorporated Mach's principle. Einstein's scheme never established a direct connection between local inertia and distant matter.

In the 1950s, British astrophysicist Dennis Sciama made a well-regarded attempt to bridge the gap. He wrote down equations designed to make the locally measured mass of a particle depend on the rest of the matter in a continuously expanding universe. According to his calculations, the enormity of material in the cosmos would outweigh disparate regional influences and produce the uniform tendencies we know as inertia. Sciama never fully developed his model, however—he passed away in 1999 before completing his grand vision. Other physicists have launched similar efforts to encompass Machian notions, but none of their schemes have panned out so far. Perhaps their imaginations haven't been properly nourished, say with cheap, wholesome cuisine.

Enter a trio of hungry cosmologists, famished for truth and a hearty meal. One of us (Paul Wesson), invited colleagues Sanjeev Seahra and Hongya Liu to a working dinner at a no-frills restaurant in Waterloo, Canada. Over heaping plates of seafood, the trio pondered ways of formulating Mach's principle in terms of gravitational waves moving through an altered version of Einsteinian space-time. Through streams of relativistic calculations, hastily jotted down on available napkins, an intriguing picture emerged of a profoundly interconnected cosmos.

The modified theory involves expressing the space-time metric (which measures distances between space-time points) in *complex* numbers, instead of real (ordinary) numbers. Complex numbers, including terms such as the square root of negative one, play little role in traditional gravitational physics. However, they comprise an important part of quantum mechanics, helping to explain hidden connections between particles. In particular, they permit a complete description of particles in terms of "wave functions": entities that can stretch out over vast regions of space.

By describing mass in terms of elongated waves rather than conventional clumps, the group found that it could express local inertia as a manifestation of the geometry of the universe as a whole. Thus, the combined effects of curvature throughout the entirety of space-time could exert a tug significant enough to affect the acceleration of objects on Earth. The more matter in space-time (such as stars, galaxies, and quasars), the greater its fabric bends and the more pronounced the effect.

This approach to Mach is compatible with Einstein's standard theory but goes considerably further. In a mathematical sense it extends general relativity to complex numbers, opening the way to all sorts of wavelike phenomena that were formerly the purview of quantum mechanics. In a physical sense the idea that a particle is a wave—whose behavior depends on the rest of the matter in the universe—links the local to the remote. This result came as a surprise to both quantum and classical physicists familiar with the approach, since it shows a way of bridging the two topics. More work is under way to see if the bridge represents a broad boulevard or just a catwalk.

Ancient mariners used to steer by the stars—relying on those distant beacons to help them sail across uncharted seas. If Mach's principle is true, the stars guided their vessels in subtler ways than they ever could have imagined.

3   Eternity in an Hour:
The Accelerating Universe

Copernicus taught us that we don't live in a special place in space. Translated into time, that led to the very important Copernican principle that all points in time are the same. Now we've discovered that the universe is accelerating, and we do live in a special place in time. We're right near the transition point between deceleration and acceleration, and not all times are the same. I think that is something that has to have profound meaning for science.

Paul Steinhardt, Princeton cosmologist

EINSTEIN'S GREATEST BLUNDER

After Einstein completed general relativity, he was satisfied but rather exhausted. The intensity of the project took a toll on his health. Nevertheless, he felt intellectually compelled to apply his master-work toward unraveling one of the deepest mysteries of science: the shape and form of the cosmos itself.

Today, the notion of galaxies as immense groupings of stars is so familiar that it's hard to believe the concept is less than a century old. Before Hubble measured the distances to Andromeda and other spiral forms in the sky in 1924 and established them as "island universes" in their own right, many astronomers thought they were simply nebulas (gas clouds) within the Milky Way itself. In other words, astronomers believed that the Milky Way constituted the entire universe and that all celestial bodies belonged to it. The cosmos, they thought, was a homogeneous sea of stars (and other formations) that had remained roughly the same since the beginning of time.

Before Hubble's discoveries, Einstein shared this early perception, believing that the overall distribution of material in space was essentially static. Therefore, when he applied general relativity toward the universe, he was astonished to discover that his result was highly unstable. Like an acrobat teetering on a wire, a slight push in any direction would send his model flying. A bit too much matter and his solution collapsed. A bit too little and it blew up. In either case, the universe seemed a fleeting creation, not a rock of the ages.

Reluctantly, the German physicist felt compelled to supplement his elegant equation with an extra term, known as the cosmological constant or the Greek letter Λ (lambda). This addition served to stabilize his model of the universe by counteracting gravitational attraction with a kind of antigravitational repulsion. It effectively offered a balancing pole to the teetering acrobat. Where the antigravity came from, Einstein couldn't say. Finding it a bit crazy, he informed his friend Paul Ehrenfest that he had "committed something in the theory of gravitation that threatens to get me interned in a lunatic asylum."

The geometry Einstein had chosen for his model of the universe too was rather unusual. Instead of a stretched-out, speckled sheet, as we often imagine the canopy of the heavens to be, it resembled a polka-dot balloon. Rather than infinite, it was closed and finite. A beam of light heading in any direction would circumnavigate the entire universe and eventually return to its starting place.

Einstein selected a bounded, rather than unlimited, cosmos purely for philosophical reasons. He ardently wanted general relativity to obey Mach's principle—with the distant stars guiding local inertia—but found that he couldn't do so for an infinite collection of stars. A finite universe would fit that model much more easily. Naturally, though, the universe couldn't end with a wall. It would be far more eloquent to imagine the cosmos as sufficiently curved that it connects up with itself—in other words, as what mathematicians call a "hypersphere."

A hypersphere is a higher-dimensional version of an ordinary sphere. Take a dot, spin it around a loop, and it becomes a circle. Twirl that circle about an axis and it becomes a sphere. Now choose an additional dimension, perpendicular to the ordinary three dimensions of space, and whirl that sphere around. It traces out a higher-dimensional object. Naturally, that last step of this extrapolation is hard to fathom. Yet there are creative ways of picturing higher dimensions and of determining the actual geometry of the universe.

THE SHAPE OF THE MATTER

Let's say that you've never heard of the game of basketball. You come from a tiny island nation where the only two sports are synchronized and unsynchronized swimming. Suppose you enter a gym in the United States and see a basketball on the floor. Without picking it up, how do you know it's spherical?

The answer is trickier than you might think. Our vision carves out two-dimensional planes in three-dimensional space. Yet nuances of shade and color, the diminution of apparent size with distance, and varied perceptions from each eye offer us a sense of depth. These optical tools help us ascertain objects' shapes and positions. Artists make use of such subtle cues to enliven their works, lending them an extra dimension. Such illusions leap out at us most vividly in 3D movies.

How then can we really be certain that a basketball is spherical, not just a cleverly disguised orange pancake? A sure way of telling involves measuring angles on its surface. If you trace out a triangle on a flat pancake and add up its angles, the sum is precisely 180 degrees. Do the same for the exterior of a ball, and you arrive at a figure greater than 180 degrees. The differences between flat and curved surfaces were first noted by the mathematicians Gauss, Lobachevsky, and Bolyai in the early 19th century, and involve the subject known as non-Euclidean geometry.

Now imagine an ant crawling along the basketball—perhaps the same savvy insect that rode on top of Newton's bucket. Constrained on the ball's surface, it would be unaware that the ball has depth. It would believe that it lives in a two-dimensional world. However, two unmistakable facts would strongly suggest that the ball is a sphere. First, the ant could easily circumnavigate the surface and return to its original position. That would at least tell it that the ball is finite, not stretched out indefinitely. Second, it could measure out its own triangle and sum up the angles. A quick calculation would prove that the surface is curved.

But suppose the ant was easily distracted and somehow never completed a full circle around the ball. If it could never physically enter the ball's interior, how could it really be sure that it lived in a three-dimensional world? Maybe it would even discount the presence of a third dimension, since it couldn't actually see it. Dismissing mysterious, unseen directions, perhaps the ant would conclude that it resided on a two-dimensional pancake with strange geometric properties. Becoming an expert in non-Euclidean geometry, it would consider the basketball's inside merely a hypothetical construct, lacking a physical basis.

Similarly, what if astronomical observations in space and time show that the four-dimensional space-time of general relativity is actually curved? Then we are led to ask: Curved into what? The logical answer is that space-time bends into the fifth dimension, which we may not be able to sample directly because we cannot step out of our world; which leads us to a fundamental question not so much technical as conceptual: Are extra dimensions merely hypothetical constructs, convenient for mathematical discourse, or are they in some sense *real?*

The nature of reality in science is a tricky game. For example, in quantum physics, particles are described by wave functions, containing all manner of information. However, at any given time not all these data can be accessed. According to Heisenberg's uncertainty principle, by taking a measurement of one quantity of a particle (position, for instance), other quantities (such as momentum) tend to blur. Hence, these measured quantities, called observables, generally don't constitute a complete picture of the particle. One might ask, then, which is the true physical reality—the shadowy realm of wave functions or the incomplete set of observables?

Astronomers face a similar dilemma when they examine phenomena that cannot be directly observed. Consider, for example, the hundreds of planets discovered, during the past decade, to be orbiting distant stars. Most of them were detected through their gravitational tugs on their suns. Assuming (as in most of the cases) that the planets themselves are too dim to be seen, astronomers must *infer* their existence. They are presumed real because that's the best explanation researchers have developed to account for their parent stars' slight movements. In our own solar system, for many years Neptune's existence was merely presumed. Well before its image was seen with a telescope, astronomers surmised its presence from perturbations in the orbits of the other planets. Did those observations alone make Neptune real, or did its light have to be detected first?

Most physicists and astronomers today would say that something is real if they infer its existence through a logical explanation that preserves the established laws of nature. The subatomic particle called the neutrino is a good example of this philosophy. Theorist Wolfgang Pauli postulated its existence through applying the principles of the conservation of energy and momentum. Although his peers gently taunted him about his advocacy of a particle that had never been seen, Pauli stood his ground. Almost two decades later, experimenters finally proved him right. The neutrino, though notoriously one of the most elusive particles in nature, was indeed real.

Einstein went back and forth throughout his career on whether or not extra dimensions were real. This indecision related to his mixed feelings about the role of experimentation in physics. Philosophically, he had one foot firmly in each of two camps. He often argued that experimentation was needed to establish any proposition. That's why he breathed easier once the Mercury precession and light-bending measurements seemed to confirm general relativity. On the other hand, he spent much of his later years trying to use his own intuition to surmise the deep mathematical principles underlying reality. At least to the outside world, these musings seemed to have little to do with what was experimentally known at the time.

Einstein's propositions that the universe is shaped like a hypersphere, and that a cosmological constant is needed to bolster it from either expansion or collapse, could not be tested for many years. Only in recent times have astronomers been able to map the likely shape of the cosmos and consider the likelihood of an antigravity term. Nevertheless, just by bringing up these issues Einstein ushered in a new age for cosmology. For the first time, science addressed the possibility that space itself has an overall shape.

A sphere is not the only way a surface can be curved. Saddles, for example, are often curved one way on the bottom, to accommodate the horse, and another way on the top, to provide comfort to the rider. Similarly, three-dimensional spaces can curve several ways into a higher dimension while preserving constant curvature. Besides a hypersphere (known as closed or positively curved), spaces can be saddled-shaped hyperboloids (known as open or negatively curved). The third possibility is for the space to be completely flat (known as zero curvature).

In 1922, Russian mathematician Alexander Friedmann explored each of these geometric possibilities for the universe. In the absence of a cosmological constant, he found that they corresponded to three distinct cosmologies. He characterized these by a parameter, known as the scale factor, that measures the size of space. If space grows, for instance, the scale factor increases over time. This results in its content (galaxies in the present era) moving farther apart.

The universe, as Friedmann envisioned, started out extremely small. Then, like a pumped-up balloon, it began to expand. If the universe's overall geometry is closed, this expansion will eventually reverse itself—like air being let out of a balloon—and collapse it back down. This catastrophic demise is often called the Big Crunch. If, in contrast, the universe is open or flat, it will expand forever. The difference between the two models pertains to how quickly the scale factor grows; it grows faster for open than flat geometries. These three possibilities (closed, open, and flat) delineate what are known as the Friedmann cosmological models.

Which geometry is feasible for a particular universe depends on its overall density. Denser universes follow a closed scenario, while sparser ones obey an open scenario. Universes of densities precisely equal to a critical value are flat. The ratio between the actual density and critical density is called the omega parameter. For omega greater than one, the cosmos is closed; less than one it is open; and equal to one it is flat.

According to physicist George Gamow, Friedmann sent his results to Einstein, pointing out inaccuracies in Einstein's static model. Einstein did not reply for quite some time. Finally, he responded with a "grumpy letter," reluctantly agreeing with Friedmann's conclusions. Although Friedmann published his results in a prestigious German journal, they were overlooked for several years—until Hubble's remarkable findings brought them to prominence.

EXPANDING PERSPECTIVES

Hubble's discovery, in 1929, of the expansion of the universe came at a fortuitous moment for cosmology. By that time, general relativity was a well-established theory with a host of solutions. In addition to Einstein's static universe and the dynamic Friedmann models, de Sitter had proposed a curious cosmological model that was completely empty but expanded anyway. Thus, scientists wishing to describe the cosmos could choose from possibilities galore: stable, dynamic, expanding, collapsing, full of matter, bereft of matter, and so on.

Consequently, when Hubble revealed that all the galaxies in creation were fleeing from each other like a roomful of angry solipsists, theorists were well prepared. They dusted off expansion scenarios and put them to good use—relegating static models to the bottom drawers of musty filing cabinets. The final vestige of the Newtonian cosmos—the notion that space doesn't evolve—crumpled under the weight of immutable facts.

It did not take long for Einstein to realize he had erred in presuming that the cosmos was immutable. In January 1931, during a trip to the United States, he visited Mount Wilson Observatory in California to see for himself the instrument that had provided a window to cosmic truth. By that time Einstein was extraordinarily famous, so film crews accompanied him as he rode the elevator up to the 100-inch Hooker telescope and glanced through its eyepiece. Hubble was beaming with pride as he showed the German physicist the most powerful telescope on Earth and the spectral evidence he had gathered with it. Paying tribute to Hubble's work, Einstein admitted that the cosmological constant had been a mistake. "Not for a moment," said Einstein, "did I doubt that this formalism was merely a makeshift to give the general principle of relativity a preliminary closed form." The purest form of the general relativistic equations, he declared, had been the correct one. The age of the expanding universe had begun.

One of the articles unearthed at that time was remarkable for its prescience. Written in 1927, it came into prominence in 1931, when Eddington had it translated from French into English and published in the *Monthly Notices of the Royal Astronomical Society*. The piece was remarkable for predicting not only a growing universe but also one whose growth is tempered or accelerated during various phases of its life. Moreover, it suggested a seeming connection between cosmological theories and the biblical notion of a moment of creation. Perhaps this was not surprising, given that the author of the paper, Georges Lemaitre, was an ordained priest as well as a scientist.

Born in Belgium in 1894, Lemaitre studied math and physics while attending seminary, devouring all he could read about general relativity. After his ordination in 1923, he attended the University of Cambridge, where he took courses under Eddington. He completed his education at MIT, obtaining his Ph.D. in 1927.

In his seminal research paper, Lemaitre devised a hybrid between the cosmological theories of Einstein and Friedmann. Adding a cosmological constant to Friedmann's disparate geometries, Lemaitre found that Einstein's equations produced a curious assortment of behaviors. The resulting solutions became known as the Friedmann-Lemaitre models.

The solution Lemaitre found the most promising is sometimes called the "hesitation universe." According to his theory, all of space and time began with a solitary burst of energy—a singular moment of genesis. Before that explosive instant, absolutely nothing existed. Afterward, the universe was a rapidly growing fireball, hurled outward by the blast. Fred Hoyle, a leading critic of this idea, later dubbed it the Big Bang. Lemaitre preferred to call the initial state the primordial atom. (It was sometimes also called the cosmic egg.)

During its initial era, the cosmos was very dense. Consequently, the sticky force of its gravity was strong enough to slow the expansion. As the universe got bigger and bigger, its expansion became slower and slower. Eventually, its expansion was languid enough that galaxies could assemble from the hot matter. The galaxies in this model were distributed like a fluid with no center and no edge. As in the Einstein universe, they resembled dots painted on the surface of a higher-dimensional "balloon." Just as every point on a balloon is as central as every other point, no galaxy can rightly claim to be in the middle of the universe.

According to Lemaitre, the lazy-growth period or "hesitation era" lasted for billions of years, allowing for the formation of all the galaxies we see in the sky. Then a new force began to dominate cosmic dynamics—the repulsive power of the cosmological constant term. We now call this extra push the "dark energy."

One of the advantages of Lemaitre's proposal was its flexibility. By tinkering with the value of the cosmological constant, one could reduce or extend the hesitation era as much as one wanted—like tuning a radio dial to produce the best reception. Presumably the optimal time frame of Lemaitre's model would be one that reproduced the known age of the universe and other observed astrophysical facts.

Although Eddington helped bring Lemaitre's paper to publication, he vehemently disagreed with its premise of a universal beginning. The British astronomer found distasteful the idea that time could have a starting gate, preferring to believe that the cosmos existed eternally. To remove the concept of genesis from the equations, Eddington pondered an infinitely long quiescent period, similar to Einstein's static realm, in which the universe was like a solid lump of dough. This space-time dough would have persisted in the same state forever, except that somehow a disturbance (acting as a kind of cosmic yeast) caused it to rise. It expanded, under the influence of a cosmological constant, until it reached its present-day size—hence the colossal cosmos we observe today.

Why would a sleeping cosmos of infinite duration suddenly wake up? This profound philosophical question dates at least as far back as St. Augustine of Hippo. In *City of God*, he argued that there was no contradiction between an immortal creator and a finite creation at a fixed instant in time. Eddington believed the awakening stemmed from a chance occurrence that could have occurred at any moment. If someone were to bet in the lottery an infinite number of times, eventually they'd win and their life would be changed forever. The universe simply won the lottery.

A third alternative, to both a singular explosion and a slow waking up, dates back (at least philosophically) to traditional Eastern notions of eternal cycles. The Hindus, Babylonians, dynastic Chinese, ancient Greeks, and many other cultures have advocated an ever-repeating universe in which the slate is periodically wiped clean. In the mid-1930s, Caltech physicist Richard Tolman explored a similar concept with his "oscillatory universe." According to this model, instead of a universal beginning, the Big Bang was preceded by the "Big Crunch" of an earlier cycle. That crunch stemmed from the earlier era's collapse, which was precipitated by a previous Big Bang, and so forth. Each era resembled a closed Friedmann model, glued by fate to its predecessors and successors.

Tolman realized, however, that his model could not produce an endless succession of viable worlds. Rather than starting afresh, each era would preserve the entropy (amount of disorder) of the previous era. Like a movie theater that never sweeps up between screenings, the universe would accumulate more and more disorderly energy. Tolman calculated that this entropy increase would make each cycle longer and longer, with higher and higher temperatures, while less and less hospitable to the development of galaxies, stars, planets, and life. Ultimately, the cosmos would recycle itself into an indefinite array of lifeless stages. We might ask a philosophical question: If a universe arises that no living being is around to observe, does it truly exist?

Note that the various cosmological theories of that period had markedly different suppositions. Both Lemaitre's model and Eddington's model made use of a cosmological constant term. Even though Einstein called this term his greatest blunder, it offered cosmologists greater freedom to "fine-tune" each universe model to bring it into line with astrophysical predictions. Tolman's model, on the other hand, based on an extrapolation of Friedmann's universe indefinitely into the past and future, did not have such a term. Hence, it was simpler but did not possess the same flexibility.

Einstein expressed interest in the variety of cosmological models that attempted to explain Hubble's discovery, but he did not step into the fray and advocate a particular scenario. He consulted with both Tolman and Lemaitre but did very little research himself on this question. (A paper Einstein published on the topic mainly summarized what was known at the time.) While this discourse was taking place, Einstein had become intensely focused on a different goal: to describe two of the known forces of nature (electromagnetism and gravitation) by means of a unified field theory that would replace quantum theory with an alternative explanation of atomic phenomena. Ironically, while Einstein pressed on with this goal, it was a third interaction—the nuclear force—that would come to dominate discussions in physics for quite some time.

Forging the Elements

By the 1940s it became clear to the astronomical community that any credible theory of the universe would need not only to have expansion but also to address the origin of the chemical elements. Hans Bethe (who passed away recently at the age of 98—still productive in his later years) had proposed a brilliant model of stellar nucleosynthesis, showing how helium, carbon, and other higher elements could be built up from hydrogen through the process of fusion. Hydrogen nuclei (protons) could weld together to form deuterium (a heavier form of hydrogen, with both a neutron and a proton in each core). Deuterium, in turn, could meld with hydrogen, to form helium-3 (two protons and one neutron per nucleus). Helium-3 nuclei could fuse into helium-4 (two protons and two neutrons per nucleus), releasing protons in the process, and so forth.

These processes could not occur just anywhere, however. They needed extremely high energies to overcome the electrical repulsion of protons—enabling these particles to be close enough to feel the attractive nuclear force. At the time of Bethe's proposal, it was unclear if stars were hot enough to produce all the higher elements in the universe (beyond helium). It was also uncertain how such material, once created, could be disseminated.

Russian physicist George Gamow, a former student of Friedmann's, found in Lemaitre's notion of a "primordial atom" a perfect opportunity to explain how the ultrahigh energies needed for nucleosynthesis could arise. Along with Ralph Alpher and Robert Herman, young researchers at Johns Hopkins, Gamow proposed that all the known elements, from hydrogen to uranium, were forged in the blazing furnace of the Big Bang. The Big Bang, they reckoned, was hot enough to allow for the assembly of dozens of elements out of hydrogen building blocks.

Gamow had a splendid sense of humor and could not resist a good joke. When in 1948 he submitted his paper, placing Alpher's name first and his own name last, he could not resist inserting Bethe's appellation in the middle. This did not reflect an actual contribution by Bethe to the project. It was just so that the "authorship" of the paper—Alpher, Bethe, and Gamow—could resemble the first three letters of the Greek alphabet: alpha, beta, and gamma. Like a mischievous schoolboy who had just pulled off a prank, Gamow sent a copy of the paper to his friend Oskar Klein. He included this personal message to Klein: "It seems that this 'alphabetical' article may represent alpha to omega of the element production. How do you like it?"

Perhaps not quite realizing the importance of the paper, Klein wrote back: "Thank you very much for sending me your charming alphabetical paper. Will you allow me, however, to have some doubt as to its representing 'the alpha to omega of the element production.' As far as gamma goes, I agree of course completely with you and that this bright beginning looks most promising indeed, but as to the further development I see difficulties." In pointing out that Gamow hadn't accounted for *all* the Greek letters, just three of them, Klein was only kidding. Ironically, there were indeed real physical difficulties with the paper, going beyond its "bright beginning," but Klein didn't find them. It took a rival group of scientists to point out some of the model's limitations.

That rival group, including British cosmologists Hermann Bondi, Thomas Gold, and later Fred Hoyle, advanced what is known as the "steady state theory." Their theory derived from profound philosophical objections to the Big Bang. Like Eddington, they couldn't imagine the cosmos emerging in a flash. Given the time-tested laws governing the conservation of matter and energy, they found it preposterous that all the material in the cosmos could suddenly arise from nothing, like a magician's trick. Why should one moment in the universe's history be so radically different from all the other moments?

Bondi and Gold proposed a new law of nature, called the "perfect cosmological principle," stating that the universe has maintained a largely consistent appearance for all times. Billions of years ago, according to this view, there were different stars and galaxies—perhaps even different life forms—but the overall distribution of these objects was roughly the same as it is now. This is an extension of what is called the "cosmological principle," the Copernican notion that Earth has no special place in space. The perfect cosmological principle further supposes that Earth has no special place in time as well.

If the cosmos has been as consistent throughout the ages as Dorian Gray's visage, how can we explain the Hubble recession of the galaxies? Doesn't universal expansion imply change? The steady state theory addresses this issue by purporting that as the galaxies move away from each other, a gradual infusion of new material would fill in the gaps, leaving everything pretty much the same. This process is called "continuous creation." The amount of new matter needed to restock the vacant regions and maintain consistency over time would be extraordinarily tiny. Just one hydrogen atom per cubic mile of space would need to materialize each year. Eventually, these newly created atoms would coalesce into new galaxies, replacing the older ones that have moved away.

Critics of the steady state theory pointed to its continuous creation of atoms as an egregious violation of the conservation of mass. How could atoms simply appear out of nowhere? Proponents of the theory responded that a minuscule violation of a physical law, spread out over the eons, was far superior to a colossal breakdown at a single point in time. If a matter is pulled out of a hat, they argued, isn't it best performed through a slow trickle than with a Big Bang?

Throughout his career, Hoyle (later joined by astronomers Geoffrey Burbidge and Jayant Narlikar) advanced every conceivable argument for various versions of steady state cosmology. He developed the machinery for a "creation field" that would explain how new material could arise from nowhere. The stretching out of space would enrich this energy pool, providing a source for new particles. (Later a very similar mechanism would be used to explain the much more popular "inflationary" model.) Technically, one chooses the average pressure of the universe to be exactly equal in value but opposite in sign to the density, for all times. In this manner, matter would be produced at just the right rate to offset the dilution caused by the expansion and keep the density constant: a steady state.

A critical byproduct of the steady state theory was the development of a viable model of how the heavier elements came to be. In 1957 Hoyle, along with E. Margaret Burbidge, Geoffrey Burbidge, and William Fowler, wrote an extraordinary paper detailing the synthesis of elements. Elements are germinated, they proposed, in the fiery bellies of stars and released in catastrophic supernova explosions. The cores of stars, they showed, have high enough temperatures to forge each atomic nucleus from simpler ones. Thus, virtually everyone around us was once embryonic material in stellar wombs. This provocative idea was later proven correct by detailed calculations performed by Donald Clayton and other researchers. The original paper (dubbed "B²FH" after the initials of its authors) became a landmark in astrophysics. Fowler (but not the others) received a Nobel Prize in 1983 for this discovery.

As astrophysicists measured the relative abundance of various elements in space and determined the energies required to synthesize these, they realized that nature used two different means of assembly. For lighter elements, such as deuterium, helium, and lithium, Gamow's fireball proposal seemed to account well for the recorded amounts. One could use the theory, for example, to calculate the amount of helium synthesized by nuclear reactions during the fireball. This figure nicely matched observed quantities. For elements heavier than lithium, however (such as carbon, oxygen, and so forth), the fireball explanation did not suffice and the supernova theory fit well. Hence, at a birthday party, while the helium in a balloon may well have been multi-billion-year-old Big Bang leftovers, the cake's ingredients were certainly more freshly made in a stellar oven.

By the early 1960s, the debate between Big Bangers and steady staters had assumed epic proportions. Without sufficient evidence supporting either position, discussions of the issue veered toward the philosophical rather than the physical. Those who liked to think of time as precious and unique tended to agree with Gamow, while those who preferred imagining it as copious and indistinguishable tended to support Hoyle. It would take a buzz from the distant past to help settle the matter.

Microwave Clues

The old Bell Labs was well known in the 1960s (and a number of years thereafter) as a haven for unfettered basic research. Though privately owned (by the phone company, no less), it kept its employees on looser reins than the government or even many academic settings.

Researchers were largely free to follow their own creative instincts as long as a reasonable number of their projects eventually bore fruit. With enough brilliant people pursuing their dreams, the result was a steady stream of groundbreaking achievement in fields from linguistics to physics.

Arno Penzias and Robert Wilson are two of Bell Labs' most famous sons. In 1965, while scanning for radio emissions from a gaseous ring surrounding the Milky Way with the giant Horn Antenna in Holmdel, New Jersey, they uncovered veiled truths about the essence of deepest space. Designed for satellite communications as part of NASA's Project Echo, the antenna assumed profound astronomical importance in the hands of these capable researchers. Unexpectedly, the funnel-shaped aluminum structure acted like an ear to the distant past. To their amazement, instead of satellite signals or more conventional reverberations, they encountered the echoes of the early universe. Their unprecedented findings demonstrated that the cosmos is bathed in the cooling afterglow of a searing earlier epoch.

Detecting and analyzing astronomical radio waves is a tricky business. There are many different types of earthly noise (such as radio and television broadcasts) that can mask celestial signals. Consequently, when Penzias and Wilson were preparing the Horn Antenna for their sky scan and heard a strange persistent background hiss, their first thoughts were to rule out a variety of mundane possibilities. Aiming the receiver in a wide range of directions, they were surprised that the background noise did not vary at all. It seemed to be coming from everywhere. As a last stab at eliminating the peculiar sound, they decided to investigate the possibility that "white dielectric material" was fouling up the receiver. You may have seen such a substance on windshields from time to time. It drops out of pigeons. But *no*, after thoroughly cleaning every square inch of the antenna, the hiss remained.

Finally, Penzias and Wilson decided to consult with Dicke, just down the road at Princeton. Dicke, as it turned out, was planning to search for relics of the Big Bang in the high-wavelength (radio and microwave) region of the spectrum. He had long suspected that hot primordial radiation, cooled over time through cosmological expansion, would be present throughout the cosmos. Along with young astronomers P. J. E. (Jim) Peebles and David Wilkinson, he was developing a radiometer to scan for such remnant signals. They were astonished to learn that they had been beaten to the punch.

The Princeton group quickly calculated the temperature of the radiation that would produce the signals that Penzias and Wilson observed. It turned out to be roughly three degrees Kelvin (three degrees above *absolute zero*, or minus 454 degrees Fahrenheit). Then, employing techniques in thermal physics, they determined the temperature of a fireball that had been chilled by billions of years of expansion. That value also turned out to be a few degrees Kelvin. Hence, Dicke and his co-workers proclaimed Penzias and Wilson's findings as proof that the universe was once enormously hot and dense. The low-temperature radiation that fills all of space became known as the cosmic microwave background (CMB).

When Penzias, Wilson, and the Princeton group published these results, they were proclaimed as the most important cosmological discovery since the time of Hubble. In a stunning omission, however, the articles did not cite key work by Gamow, Alpher, and Herman regarding the temperature and content of the early universe. Gamow hurriedly pointed out that he and his colleagues had predicted the relic radiation back in 1948. In his memoirs, Dicke later wrote:

> There is one unfortunate and embarrassing aspect of our work on the fireball radiation. We failed to make an adequate literature search and missed the more important papers of Gamow, Alpher and Herman. I must take the major blame for this, for the others in our group were too young to know these old papers. In ancient times I had heard Gamow talk at Princeton but I had remembered his model universe as cold and initially filled only with neutrons.

By the late 1960s, the steady state theory was tottering. Despite repeated attempts to amend it by Hoyle and his associates, it lost considerable support. Lacking a fireball stage, it simply could not account for the origin of the CMB. The battle seemed to have been won by the Big Bangers—at least for the moment. For their epic discovery Penzias and Wilson would receive the Nobel Prize for physics in 1978.

TIME'S BEGINNINGS

For about a decade after the CMB was discovered, the astronomical community (for the most part) stood entranced by its achievement— so awestruck by the Big Bang model that few among them pointed out any flaws. At last humankind could delve into the first instants of time and pen a new scientific Genesis. The major issue that needed to be worked out, many scientists seemed to argue, was the precise timing of cosmological events.

In 1977 a book by physicist Steven Weinberg, audaciously entitled *The First Three Minutes*, celebrated humankind's newfound ability to map the infant moments of the cosmos. It provided a remarkably detailed picture, dating as far back as one-hundredth of a second after the initial explosion. Weinberg explained:

> Throughout most of the history of modern physics and astronomy, there simply has not existed an adequate observational and theoretical foundation on which to build a history of the early universe. Now, in just the past decade, all this has changed. A theory of the early universe has become so widely accepted that astronomers often call it the "standard model."

Weinberg proceeded to explain the step-by-step process by which matter was created—from elementary particles such as photons, electrons, protons, neutrons, and neutrinos, to deuterium and then higher elements. Each successive phase occurred as the universe cooled enough to accommodate more complex structures. Yet even he admitted, "I cannot deny a feeling of unreality in writing about the first three minutes as if we really know what we are talking about." However, the unanswered questions seemed mainly to concern the *fate* of the universe rather than its origins, because researchers of the time shared a feeling that the universe could well be modeled by the dynamics prescribed by Friedmann. Recall that there are three basic Friedmann models: closed, open, and flat. These models are characterized by a parameter "omega" that relates the actual density of the universe to a critical value. If omega is greater than one, the universe is closed and doomed eventually to collapse. If, on the other hand, omega is less than or equal to one, the universe is open or flat, respectively, and fated to expand forever. Thus, the burning question of the time concerned the exact value of omega.

The omega question is akin to asking whether or not a rocket has enough impetus to clear Earth's gravity and blast off into space. If its initial thrust is piddling, there's no way it can make the jump. Rather, it will arc back down toward the ground and crash. With sufficient liftoff speed, however, its momentum will take it well past Earth's gravitational pull and deep into the interplanetary void. These two possibilities are analogous to omega greater than or less than one, respectively. A third possibility, analogous to omega equals one, is that the rocket would have just the right initial push to propel itself into orbit. It would neither be forced down by Earth's gravity nor escape it. Instead, it would be forever at the brink of conquest and freedom—allowed to sail but forbidden to follow an independent course. Such is the fate of a flat universe. What destiny then is in store for us?

Even in the 1960s and 1970s, before the advent of space telescopes and other precision instruments, astronomers knew something about omega. They realized that it was probably not miniscule or enormous (one-thousandth or 1 million, let's say) but rather stood reasonably close to one (within what scientists call an "order of magnitude" or factor of 10). This extraordinary proximity to a particular value introduced a thorny theoretical problem known as the "flatness dilemma."

As Flat as a Pancake

How special is the universe? Are its features akin to a Rolls Royce or a Yugo—meticulously assembled to order or common mass production? Western religious tradition suggests that the cosmos was custom-made for man. If it weren't for the slipping in of sin by slithering agents, we'd be living in a state of sheer perfection. Eastern belief is perhaps less egocentric, positing that our race and civilization comprise but a minute component of endless creations. Scattered through space and time, like myriad shiny pebbles on a surf-scrubbed beach, every conceivable possibility exists.

Since the age of Copernicus, science has veered steadily away from specialness. Earth, it asserts, is but an ordinary rock tucked into an average corner of the cosmos—a speck in the dustbin of the utility closet of space's arena. Life is but a random brew, concocted by the blind chefs of time. Consciousness is merely a curious combination of chemicals whose interactions cause awareness. Art and poetry stem from particular firings of neurons that trigger pleasing receptors in the brain—and so forth.

Perhaps the ultimate expression of this philosophy is the so-called "chaotic cosmology programme"—a phrase coined by cosmologist John Barrow to characterize a far-reaching scientific goal. Given the complete range of possible characteristics of the early universe, it says, which of these could have resulted in the current cosmic state of affairs? Barrow designated the complete set of initial possibilities to be the collection of all possible solutions to Einstein's equations of general relativity. More recently, astrophysicist Max Tegmark has extended this to include all conceivable laws of nature.

Speculative writers and historians often consider such "what if" scenarios—applied to Earth, that is. If the plague hadn't decimated Europe (or, in the other extreme, if it had left but a small percentage of survivors), would there still have been a Renaissance? Or would medieval institutions have lingered for many more centuries— possibly even until the present day? Just as one might contemplate alternative histories of Earth, one might consider disparate scenarios for the universe itself.

How wide a range, for example, could the value of omega have been in the earliest stages of the universe and still lead to the current state of affairs? If omega started out as one-half (representing an open universe) or two (representing a closed universe), to pick some values, could the cosmos have evolved over billions of years into present-day conditions? Or to put this question another way: If a cosmic designer threw a dart to select the initial value of omega, how close to bull's eye would it need to land?

The answer, according to theorists' calculations, is astonishing. If, by the end of the first second after the Big Bang, omega differed from one by as little as one part in one quadrillion (the digit one, followed by 15 zeros), this minute discrepancy would have ballooned over time. In the eons that followed, the ensuing dynamics would little resemble that of the actual cosmos. Omega, by now, would be either much too large or way too small. That is, if the universe wasn't extraordinarily close to flat to begin with, it could not possibly be anywhere close to flat today. (Note that "flat" in this context refers to the shape of the ordinary spatial part of four-dimensional space-time.) This conundrum, first posed by Dicke in the late 1960s, is called the "flatness problem."

To picture this bizarre situation, imagine if the Three Bears opened a bed and breakfast and Goldilocks was one of their customers. Upon arriving at the inn, she found that her bed had no linens on it. Next to it, on top of a table, was a folded pile of sheets. A sign hung above it: "We are currently hibernating and shouldn't be disturbed. Please help yourself to a sheet. Note that these are enchanted sheets and must be placed perfectly flat upon the bed."

Goldilocks was very sleepy. Ignoring the sign's warnings, she picked up a sheet and placed it loosely on the mattress. "Flat enough," she thought, and then fell asleep. Imagine her horror when an hour later she woke up completely entangled in the sheet. It had curled up and was starting to squeeze tighter and tighter. "This sheet is too snug!" she screamed as she pulled herself away.

As soon as Goldilocks left the bed, the sheet folded itself and then hopped back up on the table. She decided to try again. "This time I'll make it really flat," she muttered to herself. "Great idea," the sheet echoed back. Goldilocks picked it up, placed it back on the bed, and then gently tucked in its corners. Trying to smooth it out, she failed to notice that one of the previous guests (a princess) had left a pea under the mattress. This tiny legume caused a minute bump in the fabric, so subtle that it could scarcely be noticed.

Nevertheless, barely an hour's time after she fell asleep again, Goldilocks woke up feeling quite odd. Suddenly, she realized that she was floating close to the ceiling. The sheet, not being perfectly flat, had billowed outward, becoming puffier and puffier until it lifted off the bed. "This sheet is too bloated!" she cried out. Cautiously, Goldilocks climbed down the sheet back to the floor. Once again, the sheet folded itself up and resumed its perch on the table.

Now, being a savvy girl, she lifted the mattress to find the source of the problem. Discovering the pea, she chucked it out the window into the garden (where it promptly grew into an ornamental stalk). Then she replaced the mattress, carefully making sure it was *absolutely* flat. Next, she spread the sheet out, eyeing it from every direction to be certain it showed not the slightest slant or kink. Once she completed her inspection, she decided to chance slumber once more. This time she was successful—the sheet was flat enough to stay that way throughout the night. Waking up, feeling the most rested she had ever been, she exclaimed, "This sheet is just right," and then checked off five happy stars on her Zagat's survey.

Just like Goldilocks, we want conditions in our world to be just right. If omega diminishes or bellows, that corresponds respectively to the universe bursting outward either too quickly or too slowly (the smaller the omega, the lesser the cosmic density and the greater it can expand). In the former case, stable structures such as galaxies would not be able to form. In the latter, the universe would expand for a relatively brief period, run out of steam, and then collapse back down to an ultradense state. Either way, Earth would not have been able to form. Thus, life as we know it is predicated on the universe starting out as flat as a Kansas cornfield.

Why should the cosmos be so flat in the beginning? Could it be that flatness is an inherent feature of the universe? Perhaps. But making such a special assumption seems most at odds with the idea that early conditions were a chaotic jumble. To resolve this contradiction, one might imagine a way of stretching out the rumpled bedsheets of the universe and eliminating all its wrinkles. Such a cosmological process would not only help ameliorate the flatness issue, it would also address another thorny dilemma, the "horizon problem."

THE WELL-TEMPERED COSMOS

When astronomers point their telescopes in any direction and map out the average distribution of galaxies, they find an astonishing degree of uniformity. Detailed galaxy counts yield essentially the same values no matter in which quadrant they are taken. Statistically, for instance, the northern sky looks virtually the same as the southern. Researchers recognize, however, that this approximation holds only for the largest scales of viewing. A more focused look reveals that many galaxies are actually in clumps, such as groups, clusters, and superclusters. Moreover, certain segments of the sky, the voids, are relatively empty, and other regions appear like the surfaces of bubbles. There is even a long sheet of galaxies stretching out hundreds of millions of light-years, known as the "Great Wall." The existence of these structures implies a certain degree of irregularity on smaller scales.

Nevertheless, the greater the scope of sky surveys, the smoother the picture of the cosmos they reveal. The superclusters, for instance, appear to be randomly distributed. Over the largest distances we can probe with telescopes, their density has only small fluctuations. Moreover, on the greatest scales, each sector of space has roughly the same temperature and composition. This smoothness can similarly be seen in statistical averages of the microwave background. Astronomers call this situation "isotropic," meaning the same in all directions.

Given the notion that everything emerged from the same fireball, 13.7 billion years ago, is such uniformity surprising? Indeed, it is, considering that according to the Big Bang model, light was not always free to move throughout space. Models of universal evolution indicate that photons bounced from particle to particle in a cosmic pinball game for about 300,000 years after the initial burst. Only then did atoms form, in the process known as "recombination." (This is a misnomer, since atoms were never really together in the first place.) Atomic nuclei grabbed up available electrons, leaving photons free to move through space. While the atoms would eventually coalesce into the seeds of stellar and other material, the photons would cool over time and form the basis of the observed microwave background radiation. Thus, it was during the era of recombination, not during the initial blast, that the universe's profile established itself.

The trouble is that by that era different parts of the universe had theoretically long been out of contact with each other. Various regions of space lay well beyond each other's "horizons"—the maximum reach of light (and all other forms of communication) during a particular time interval. Therefore, there should be no reason to expect the temperature of the fireball radiation to be the same in all directions. Yet the observed cosmic microwave background *does* have nearly the same temperature throughout the heavenly dome. Current data indicate that temperatures fluctuate only a few parts in 100,000. Physicists call this situation the horizon problem.

Imagine that 100 high school alumni arrived at their 10-year reunion each clad in ruffled purple dresses or suits and that you found out that the classmates had been completely out of touch for the entire decade. No phone calls, e-mails or letters had been exchanged, except to announce the time and place of the event. How would you explain such a startling wardrobe synchronicity?

You could chalk it up to pure coincidence or shared fashion sense. Or if you did some detective work, perhaps you might discover hidden commonalities that led to such color uniformity. For example, maybe a mixer was held shortly before graduation that brought all the seniors together. Suppose the sponsors of the event asked students to dress like the pop star Prince, whose favorite color is purple. At the mixer, students shared smiles and came to associate their outfits with graduation. Therefore, even when they were beyond communication for years, they retained certain commonalities.

Different parts of the universe have been out of touch for far longer than that. Nevertheless, they are all costumed the same. Could there have been some kind of cosmic "mixer" well before the photons "graduated" and moved away?

In 1969, physicist Charles Misner of the University of Maryland proposed the Mixmaster universe as a potential way of resolving the horizon problem. The Mixmaster universe is an anisotropic variation of the Big Bang theory. In the standard Big Bang, the universe bursts forth at equal rates in all directions, like an evenly gushing fountain. The Mixmaster universe, on the other hand, behaves far more erratically. It expands in certain directions while contracting in others. Furthermore, the directions of expansion and contraction keep changing in an essentially unpredictable manner. Misner believed that this chaotic behavior would help smooth out the universe, like the action of a blender, and explain why it currently appears pretty much uniform in all directions. He dubbed it "Mixmaster" after a vegetable processor advertised heavily at the time.

Subsequent research, however, showed that the Mixmaster wasn't quite as effective as first thought. Like broken thermostats in a massive apartment complex, it didn't level off the temperatures in different spaces enough. Some sectors would be hotter and others colder—unlike what astronomers observe today. Furthermore, an important paper by Stephen Hawking and C. B. (Barry) Collins, published in 1973, placed a damper on ideas that the universe ever was less isotropic than it is now. Entitled, appropriately enough, "Why Is the Universe Isotropic?," the article made a strong case that the chances of any anisotropic universe (such as the Mixmaster model) evolving into what we currently observe were effectively zero.

The authors of the paper suggested one way of handling the matter, an argument known as the anthropic principle. Established by Brandon Carter, the anthropic principle asserts that the universe is the way it is because if it were any different it couldn't possibility support advanced life. Conditions in an anisotropic cosmos would be too nasty and brutish to allow reasonable planetary systems to form. If there were any deviation from flatness and isotropy, there wouldn't be cognizant beings, and no one would live to tell the tale. It's the same reason why no one has written a book called "True Tales from the Earth's Core." No one lives in the core and no one could write such a book. Hence, by anthropic reasoning, we all live on the planetary surface.

Many physicists have dismissed the anthropic argument as too philosophical, even too religious, in nature—a far cry from the careful results of calculations. It smacks, they feel, of the line in Candide: "All's for the best in this best of all possible worlds." At least one physicist, Max Dresden of Stony Brook, jokingly attributed it to a modern-day version of anglocentrism. Just as the Victorians thought that England was the most civilized of all places, he commented, anthropic reasoning purports that our universe is the most civilized of all possibilities.

Other scientists, such as Roger Penrose, suggested a more mathematical way of constraining the initial state of the universe to be isotropic. In the "Weyl curvature hypothesis," he proposed that a segment of the Riemann tensor, called the Weyl tensor (after Hermann Weyl), must be zero at the beginning of time. A zero Weyl tensor is tantamount to pure isotropy. However, his and other people's erudite arguments would soon be overwhelmed by a simple device.

It would be Russians and Americans, still vying in the Cold War, who would arrive at a forceful solution. "Just blow it up," members of these nuclear superpowers proposed. No civilized selection process would be needed if the universe once underwent a period of ultrarapid expansion, much faster than the initial blast of the Big Bang. Everything would simply be evened out, like a forest after a tornado.

BLOWN OUT OF PROPORTION

Like many aspects of science, the origins of the inflationary model of the universe are somewhat complex. In 1981 physicist Alan Guth proposed the term "inflation" to describe an early period of extremely rapid expansion. His goal was to help resolve the flatness, horizon, and other problems plaguing the standard Big Bang model. Thus the scientific community considers him the father of inflation.

Largely unknown to the West at that time, however, Russian scientists had developed aspects of this notion even earlier. In the early 1970s, Andrei Linde and David Kirzhnits, of Moscow's prestigious Lebedev Physical Institute, first investigated the cosmological consequences of symmetry breaking in the very early universe. Symmetry breaking is a transformation in particle physics that creates a favored direction, akin to a spinning top falling over to one side. These ideas led Linde and Gennady Chibisov to explore the implications of such changes in the vacuum of space and inspired Alexei Starobinsky to propose a theory independently very much like inflation.

When Guth started thinking about the maladies ailing the standard Big Bang model, he had little background in cosmology and was unaware of these alternative theories. Born in 1947 in the town of New Brunswick, New Jersey, to a family with a small grocery store, Guth became interested in science when in high school. An astronomy book inspired him to contemplate possible beginnings of the universe. Nevertheless, when he began his undergraduate studies at MIT, he decided to work with Aaron Bernstein, an experimental nuclear physicist. Guth stayed at MIT to receive his Ph.D. and then began research fellowships at Columbia, Cornell, and Stanford.

It was a talk by Dicke at Cornell that changed Guth's career path, inspiring him to revisit his youthful interest. As Guth remembered it:

> One of the things he talked about was the flatness problem. . . . The problem was that if you looked at the universe one second after the Big Bang, the expansion rate had to be exactly what it was to an accuracy of about one part in 10^{14}, or else the universe would have either flown apart without ever forming galaxies or quickly recollapsed. . . . At the time I didn't even understand how to derive that fact, but I believed it and was startled by it.

Shortly thereafter, while working with researcher Henry Tye on the problem of magnetic monopoles (hypothetical magnets with only single poles), Guth discovered a mechanism in field theory that would cause the fabric of the universe to stretch by at least the gargantuan factor of 10^{25} (one followed by 25 zeros) in the exceedingly brief interval of 10^{-30} seconds. He realized that this ultrarapid expansion would offer a credible solution to the flatness dilemma. Thus, cosmological inflation was born.

Linde followed these developments with intense interest. Born in Moscow in 1948, he came to physics by way of philosophy. In particular, ancient Indian notions of an endless succession of worlds fascinated him. This abstract curiosity about an eternal universe led him to explore the tangible realm of physics, where he quickly became adept at the formalism. Soon after Guth published his paper, "Inflationary Universe: A Possible Solution to the Horizon and Flatness Problems," Linde published his own work, entitled, "A New Inflationary Universe Scenario: A Possible Solution of the Horizon, Flatness, Homogeneity, Isotropy and Primordial Monopole Problems." Linde's paper addressed some of the issues raised by Guth. Guth acknowledged the importance of Linde's contributions and later would write that "he was generous in giving credit to my work."

All these papers centered on the idea of a phase transition from one particle state to another in the nascent instants of the cosmos. Everybody is familiar with certain kinds of phase change—water crystallizing into ice, for example. As you lower the temperature of a glass of water, eventually it locks into certain patterns. Many properties of liquid water and solid ice are very different. Blocks of ice, for instance, are less dense than cold water and thus can float.

The type of phase transition that could have happened in the very early universe is more complex in origin than familiar processes such as freezing, melting, and boiling, but shares some of the same characteristics. At very high temperatures, it is believed, elementary particles interact according to laws based on certain symmetry principles. If the temperature declines, the relevant symmetry alters, and the fluid comprised of the particles undergoes a phase change. This phenomenon, called "symmetry breaking," raises the intriguing possibility that the primordial cosmos experienced one or more phase changes as it expanded and cooled. During such a phase change, bubbles of the new phase might have appeared in the old phase, like steam bubbles in boiling water.

In the original Guth model, the current epoch of the universe originated in an amalgam of such bubbles, spawned during an earlier stage. However, this hypothesis proved awkward because it led to walls between the regions represented by different bubbles. These walls would lock up tremendous quantities of energy for which there would be no natural means of release. This was called the "graceful exit problem." Astronomers have never detected such bubble walls. Consequently, the new inflationary scenario proposed that all we see around us emerged from a *single* bubble, created during a phase transition a tiny fraction of a second after the initial burst.

The boundaries of the great bubble would mark the limit between two different types of spatial vacuum. Given that "vacuum" means emptiness, you might wonder how there could be different varieties. Modern field theory postulates that vacuum regions are not bleak deserts but rather ponds brimming with particle activity. The uncertainty principle allows for the creation of "virtual particles" that leap from the waters temporarily before rejoining the sea. The level of such activity depends upon the vacuum's overall temperature. Sometimes, like ice and water at zero degrees Celsius, two different vacuum phases can coexist, each with distinct properties. In the case of the new inflationary model, these are called the "false vacuum" (surrounding the bubble) and the "true vacuum" (inside the bubble).

Within the great bubble, the dynamics of the universe would take on an explosive pace because the false vacuum would be unstable and would want to turn into a true vacuum as quickly as possible. In less than a millionth of a trillionth of a trillionth of a second, the bubble of true vacuum would increase in volume exponentially, until the universe blew up into something like the size of a baseball. Finally, colossal stores of energy would be released from the vacuum, and the universe would revert to the far slower process of conventional Hubble expansion. From that point on, the inflationary picture matches the standard Big Bang scenario.

Think of an inflationary burst in terms of price wars among gas stations. Suppose all gas stations in a city charged five dollars per gallon. Customers would stoically go to the station nearest them and pay the high cost. However, imagine that they all found out about one station charging only two dollars per gallon. Many of them would flock to that station. This situation could spur stations in the immediate vicinity to lower their prices, then stations near those, and so forth. Very rapidly, the bubble of discount prices could well expand to encompass the whole city, vanquishing the domain of high prices.

Similarly, in the inflationary picture, a region initially so small that it was causally connected (that is, contained within its own horizon) would expand so rapidly that it would become the entire visible universe. Since all parts of the initially tiny region would be within communication's range, they'd exchange photons and thus reach a level temperature (like the temperature consistency of the human body). Then, as the great bubble blew up, regions of similar temperature and other properties would be thrust far away from each other. Once ordinary expansion kicked in, these similarities would be preserved, resulting in large-scale spatial uniformity, thereby solving the horizon problem of why widely separated parts of space are so similar.

The inflationary picture would also resolve the flatness problem by smoothing out all the wrinkles of space's fabric. Any bumps or indentations would be stretched out so quickly they'd be indiscernible. It would be like taking a rubber sheet and attaching it to trucks facing north, south, east, and west. As the trucks pulled away from each other, the surface (assuming it didn't break) would become absolutely taut. Similarly, the cosmos would become perfectly flat, represented by an omega value of one. Such absolute flatness is one of the key predictions of inflation.

Another welcome product of inflation is the magnification of minute quantum fluctuations. Ironically, although ripples in the very early universe would be stretched out, new perturbations would emerge through the random actions of quantum physics. These pockets of energy would grow rapidly during the inflationary period, attract matter through their gravitational effects, and eventually form the seeds of structure in the universe (galaxies, clusters, etc.). Hence, inflation is not just a way of justifying the large-scale smoothness of the cosmos; it also explains the universe's smaller-scale diversity.

FINE-TUNING

At the same time Linde was developing the new inflationary universe, a young physicist from the University of Pennsylvania, Paul Steinhardt, along with his graduate student Andreas Albrecht, proposed an independent version of the same theory. Like Linde's version, it avoided the graceful exit problem. Steinhardt and Albrecht joined Guth and Linde in pointing the way for a radical new conception of the early universe.

Currently at Princeton, Steinhardt has moved around quite a bit—from field to field (he's also a renowned expert in quasicrystals) and from place to place. He has vivid memories of growing up as an "Air Force brat"—relocating from base to base every three years. When he was in fourth grade, his family settled in Miami. Around that time he began to nurture a fledgling fascination for science. "Astronomy was my first interest," Steinhardt recalled. "Then I dropped it for many years. Many of the first books I read were on astronomy. That was really fascinating to me. But then I got interested in other things. I always liked science in general, as far back as I can remember."

"I had a telescope, a chemistry set, a biology lab and did physics experiments. Anything that was scientific I was interested in. Doing astronomy in Miami was difficult, because you either had to go where the lights were or where the mosquitoes were. I remember going out to the Everglades, literally running out of the car, setting up the telescope, running back to the car, then putting all kinds of stuff on myself trying to fight the mosquitoes."

During much of the 1980s and early 1990s, along with various collaborators, Steinhardt tried to perfect inflationary cosmology (henceforth we will use the term "inflation" to refer to all variations, not just Guth's original model). Many issues remained unsolved and could only be tested through astronomical data. But such information wasn't available—not just yet. For example, researchers didn't know which particular phase transition in the infant universe triggered inflation. The universe could go through several such jumps, as unified forces broke down into their constituents. Many theorists believe that the universe initially had just one type of interaction—an amalgamation of the four natural forces. Such a state represented the maximum possible symmetry, or sameness. Then, somehow gravity separated from the other three. Next, the strong force pulled away, and finally the electroweak interaction divided into electromagnetism and the weak force. The end product was the current quartet of forces. In each transition a type of symmetry spontaneously broke down.

For technical reasons, spontaneous symmetric breaking requires a type of scalar field, called a Higgs boson. Mathematical fields are classified as scalar, vector, tensor, or other categories depending on their behavior on transforming the coordinate system. In physics, scalar fields represent particles, such as the Higgs, with unique physical properties, particularly a zero value of what is called total spin and invariance under certain coordinate transformations. During a phase transition this boson undergoes a change in potential energy, akin to the plunging of a barrel over a waterfall. As it plummets, it cedes mass to one or more hitherto massless exchange particles. Exchange particles are intermediaries of the natural forces. For example, the W and Z bosons, volleyed about by other particles like balls in a tennis match, generate the weak force. Once exchange particles have mass, the forces associated with them change character, decreasing in range. Consequently the weak force represents a short-range interaction.

The mix of particles during a particular phase of the cosmos affects the dynamics of its expansion. The contents of the universe during any given period can be described as a fluid with certain physical features. Each type of fluid has an "equation of state" designating the precise relationship between its density and pressure. The form of this equation sets the universe's rate of growth during that era.

For extremely early times, we are not sure of the equation of state. However, the processes that the universe could have undergone—standard Big Bang expansion, a brief era of ultrarapid inflation, then a slowing down of growth to the current rate—all delineate possibilities for the fluid dynamics during that interval. By working together, particle physicists and cosmologists can attempt to piece together the puzzle of the dynamics of the very early universe. For instance, we can model inflation using a classical description of a fluid if the pressure is assumed to be negative. This model leads to an equation of state in which, in contrast to ordinary matter or radiation, the pressure is proportional to *minus* the energy density. (Normal materials have pressures that are fractions of a positive-valued density.) Under such circumstances, Einstein's equations of general relativity mandate exponential growth. Physicists refer to a field that could do this as an "inflaton."

Imagine a filled balloon surrounded by air. If the air pressure outside the balloon is slowly lowered, the balloon will expand. If it decreases enough it will approach zero—a classical vacuum. If it could be lowered even further, into the negative zone, imagine how large the balloon would get (assuming that it didn't pop). Thus, negative pressure would cause inflation.

Steinhardt and his co-workers worked closely with field theorists to try to match inflationary dynamics to models of particle creation and structure formation in the early universe. This scheme had to be "fine-tuned" to a certain extent. If the inflationary epoch was too short, the cosmos would not have had enough time to smooth out.

If, on the other hand, it was too long, there would be little sign of structure today. These delimiters still allowed for a number of options that, researchers hoped, more data would whittle down. Steinhardt joked that constructing the right inflationary model was like picking from a Chinese menu—selecting one item from column A, one from column B, and so on. The approach that led to the correct solution would have very particular properties and would match up precise models of the microscopic and macroscopic worlds.

Linde, on the other hand, was convinced that inflation was a ubiquitous and natural phenomenon, akin to Darwinian evolution in biology. All one needed was the *tabula rasa* of empty space. On this blank pad the quill of quantum randomness would sketch fluctuations of various sizes. Through pure chance, at least one of these fluctuations would produce a scalar field able to spark the fuse of an inflationary blast. That region of space would expand exponentially, thereby dominating less explosive sectors. As it blew up, it would produce the familiar byproducts of inflation—flatness, correlations between remote domains, and so forth. Because of its reliance on sheer randomness, Linde dubbed his model "chaotic inflation."

Steinhardt, Linde, and their various collaborators spent much of the 1980s and early 1990s developing alternative models of inflation. Many others joined in on the quest. Like the makers of Coke and Pepsi, each research team produced various concoctions, hoping that one of them would pass the taste test of astronomical inquiry. Some of the models *du jour* developed by various groups included extended inflation (where a field interacts with gravity, causing it to eventually put the brakes on inflation), hyperextended inflation, power law inflation, natural inflation, hybrid inflation, eternal inflation, and so on. Each posited a distinct mechanism for inducing, then halting, a burst of ultrarapid expansion. An exponential proliferation of papers filled the science journals to the bursting point—leading to a curvature dilemma for the flimsy shelves in researchers' offices. Physicists eagerly awaited experimental data that would help distinguish these approaches—and confirm or disprove the theory itself.

One critical source of information would be provided by new observations of the CMB's profile. From the time of Penzias and Wilson until the end of the 1980s, the only features known about that radiation were its average temperature, spectral distribution, and overall isotropy. Experiments also accounted for a small discrepancy in the radiation's temperature in opposite directions of the sky due to Doppler shifting caused by the Milky Way's motion through space. Researchers, though, believed that a more precise examination would yield evidence for the seeds of structure formation in the universe. These seeds would be minute anisotropies due to slight differences in the early distribution of matter. Because inflationary theorists aspired to explain the process of structure formation in terms of stretched-out quantum fluctuations, they hoped such anisotropies would soon be found. Conversely, if no such bumps existed, advocates of any variation of the Big Bang model would have a hard time explaining how galaxies and other structures emerged.

CRINKLES IN THE FABRIC

On November 18, 1989, NASA launched the Cosmic Background Explorer (COBE) satellite, designed to take an unprecedented look at the primordial radiation bathing the cosmos. It carried several instruments, including a differential microwave radiometer, able to discern anisotropies in the background spectrum as tiny as a few parts per million. A team led by George Smoot of Lawrence Berkeley Laboratories (LBL) analyzed and interpreted the data. Physicists nervously awaited the experiment's results. Would it confirm the Big Bang picture of a fiery beginning? Would it detect the minute raisins in the tapioca pudding of uniformity?

Tension mounted as months passed by with no wrinkles to be

found. Smoot urged patience, realizing that it would take some time for the data collected by the satellite to reach levels of statistical significance. Even after some signs of anisotropy appeared, about a year after the launch, he was reluctant to publish any results until he was absolutely certain they were valid.

During this wait, some science journalists stirred up a storm with dire warnings that the Big Bang theory was in jeopardy. A *Sky and Telescope* news item inquired, "The Big Bang: Dead or Alive?" A popular book by the physicist Eric J. Lerner, revising an outdated plasma cosmology, proclaimed in its title that *The Big Bang Never Happened.*

Steady state theorists waited in the wings, eager to come to the rescue with alternative hypotheses. In an ironic twist, Narlikar and Burbidge each pointed to the smoothness of the microwave background as evidence *against* the Big Bang theory. Along with Hoyle they developed what came to be known as quasi-steady state cosmology. Unlike the original model, it predicted an isotropic radiation spectrum—albeit produced in "mini big bangs" rather than a single explosion. Few mainstream cosmologists, however, rallied to their cause.

Finally, on April 23, 1992, Smoot enthralled scientists at a meeting of the American Physical Society with his long-awaited announcement of success. He had kept his results top-secret until the very end, checking and double-checking them to eliminate ambiguities. At one point he had even flown to Antarctica (where the cold night sky is especially clear) to take extra measurements. By the time he stood on the podium he was confident that his team had recorded the stunning visage of the early universe.

The wrinkles that the COBE group found matched up beautifully with the concept that galaxies were seeded in the early universe. Corresponding to slightly hotter or colder regions of the background, the COBE picture identified primordial zones of greater or lesser density. The denser areas constituted the kernels of cosmic structure. Nevertheless, these results still weren't precise enough to nail down key cosmological parameters and distinguish particular early-universe models (such as various inflationary scenarios). Consequently, astronomers began planning in earnest for a more detailed probe.

Meanwhile, other researchers at LBL and elsewhere pursued a wholly different way of investigating the early universe. Their investigations of distant supernovas would soon jolt the field of cosmology.

CLOCKING THE SUPERNOVAS

Saul Perlmutter, leader of the Supernova Cosmology Project, grew up in a family of respected academicians. His father, Daniel, was a professor of chemical engineering at the University of Pennsylvania, and his mother, Felice, was a professor of social administration at Temple University. Nurtured in a supportive, intellectual household, his interests turned to science at an early age. As a child, recalled Perlmutter, "I always enjoyed looking at the sky, but I was never one of those people who had their own backyard telescope. It was only because I started needing telescopes to answer the fundamental questions that I started learning much about astronomy."

In addition to his scientific talents, Perlmutter became adept at music—corroborating popular theories that the two abilities go hand in hand. He's an avid violinist and enjoys playing in orchestras. Blending his talents with others—whether harmonizing in music or collaborating in science—has become an important part of his personal philosophy. "I was somebody who had fewer individual heroes and more collective heroes," he stated. "The idea that people working together could understand the world and that no single one of them by themselves could understand the world, that really captured my imagination."

After receiving a Ph.D. from the University of California at Berkeley in 1986, he was appointed to a position at LBL. Along with an international team of astronomers, including Berkeley astronomers Carl Pennypacker and Gerson Goldhaber, he set out to measure the overall dynamics of the universe and the change in its expansion rate over time. This measurement would provide a way of delving into the past and predicting the fate of the cosmos. To perform this task, they developed powerful techniques to measure the energy output of Type Ia supernovas in extremely remote galaxies.

Type Ia supernovas, the catastrophic explosions of a certain kind of star, are valuable to astronomers because they can serve as "standard candles." Standard candles are objects with well-known energy output. Imagine walking along a dark desert road and seeing a faint street lamp way off in the distance. If you know the intrinsic power of the lamp, you can deduce from its dimness how far away it is. Type Ia supernovas serve a similar purpose for astronomers eager to map the scale of the universe. By matching the apparent brightness of such stellar blasts to their actual luminosities, astronomers can reliably ascertain their distances. Light curves, indicating the progression of each burst, offer added information. Thus, they are solid celestial yardsticks, useful for measuring the remoteness of the galaxies in which they are situated.

Once astronomers know the distances to the galaxies in a given region of space, they can readily determine the expansion rate of that region. Each galaxy's spectral lines are shifted by the Doppler effect. By measuring this shift, they can assess the galaxies' velocities. Finally, by combining this information with the distance data, they can calculate how fast each part of the universe is pulling away.

In astronomy the farther out you look, the deeper into the past you see. Therefore, Perlmutter and his colleagues realized they had the perfect tool for determining how the cosmological expansion rate has altered over the eons. This tool could allow them to assess omega, the universe's density parameter, and help them decide how much of its dynamics is driven by visible material versus dark matter. A second team, led by Brian Schmidt of Australian National University and Nicholas Suntzeff of Cerro Tololo Inter-American Observatory in Chile, enacted an independent program with a similar purpose. Throughout the 1990s the two groups jockeyed for valuable telescope time and competed in a race for publications.

One aspect of the cosmos that the researchers did not expect to challenge was its deceleration. The simplest Friedmann models portrayed a universe slowing down with time. The only difference lay in how quickly this braking would occur—with the closed model representing the extreme. Adding a cosmological constant would change the situation, allowing for the option of speeding up, but few physicists saw a point to doing that. After all, even Einstein had discarded the term.

Supernova mapping is an arduous process, given that they are rare and unpredictable. It's like knocking on doors all around the country hoping to find a family with quintuplets who had just won the lottery. If observers anywhere in the world spot a distant burst, researchers everywhere must leap into action. They may need to redirect a telescope to track the supernova's light curve, with no time to spare. Then they can use that information to plot just one more point on their charts. As data trickle in, statistical significance builds over many years.

In 1998 each supernova team felt it had enough evidence to render a verdict. In startling announcements the groups proclaimed that the universe is not currently decelerating at all but rather is speeding up. Thus, not only will the cosmos expand forever, its expansion is accelerating. Each remote galaxy is moving farther and faster away from the others, with no end in sight.

DARK ENERGY

By the end of the 20th century, scientists realized that many previous assumptions about the behavior of the cosmos were dead wrong. Along with the COBE data, the supernova results pointed to a flat universe. However, unlike the simplest flat Friedmann model, with omega equal to one, the parameter associated with the expansion was only about three-tenths. In other words, the universe had approximately 30 percent of the material density associated with flat cosmologies. Something else must be hidden in the blackness of space. That extra component was named dark energy.

The simplest way of incorporating dark energy into cosmology is to reinstate the cosmological constant term, also known by the Greek letter Λ (lambda). Although that makes general relativity less elegant, it also makes it more accurate. A mathematical clarification is one thing, but a true physical explanation is another. Theorists scrambled to try to explain the origins of cosmological antigravity.

It would be incorrect, though, to picture the universe as always speeding up in its expansion. Additional supernova measurements by Schmidt's group and Perlmutter's group have revealed that the universe began to accelerate relatively recently in its history—within a few billion years of the present day. Before then the universe was dense enough so that matter terms dominated the cosmological constant term. The attraction of gravity overpowered the repulsion of lambda, slowing the expansion. Therefore, space was decelerating before it began to accelerate. Only when the universe's matter was dispersed enough did lambda begin to dominate and the universe start to speed up.

As Steinhardt has pointed out, the outstanding coincidence that we live within a few billion years of the turnaround time of the universe seems to contradict the Copernican ideal that humankind occupies no special place or time. Thus, the new results cry out for a wholesale rethinking of our basic concept of the universe. As he has remarked:

> I think people are really missing the boat on this. This is truly a revolution of Copernican nature; this is not just another addition. What the cosmology community has done for the most part is say, "Oops, we're missing an ingredient. Let's add that ingredient. Everything fits beautifully. We have a wonderful model." My reaction is: time to step back and reevaluate.
>
> The full extent of the implications hasn't been worked out yet. If

you were around at the time of Copernicus you might have said, "He wants to make the Sun the center. You want to make the Earth the center. It doesn't mean too much." But then by the time you get to Kepler and Newton it means a lot. So it wasn't just another detail. I can imagine this will be a very profound thing by the time it's through.

Steinhardt has proposed that the dark energy is a hitherto-unknown substance, called "quintessence." Its name hearkens back to the ancient notion of four natural elements—earth, air, water, and fire. Quintessence would be the fifth. Instead of a steady cosmological constant, it would be a field that kicked in during a particular epoch of the universe, causing a far milder version of an inflationary burst. Using a variable field offers greater flexibility in modeling different cosmic phases. However, current observations have not been able to distinguish between variable and constant forms of dark energy.

To resolve these and other vital issues, astronomers have pressed on with further testing. The supernova teams have continued their endeavors, accumulating a bevy of examples to enhance their data. The LBL group has proposed a space-based mission, called the Supernova Acceleration Probe, to improve their capability by 20-fold. Meanwhile, spectacular new results from the Wilkinson Microwave Anisotropy Probe (WMAP), launched in 2001, have uncovered a treasure trove of critical information about the young cosmos.

Portrait of the Cosmos as a Young Expanse

The beginning of the 21st century has witnessed cosmology becoming an exacting enterprise, with ample tools to elucidate the state of the observable universe. It has also ushered in considerable confusion as to the future direction of the field. A snapshot of the early universe captures this mixture of profound new knowledge and grave uncertainty. The stunning "baby picture" of the cosmos produced by WMAP represents one of the landmark scientific images of our times—akin to the double helix or the first photos of the Martian surface. When represented in color, like a weather map of hotter and colder sites, it is a fantastically intricate mosaic of multihued spots. Clearly the background radiation's artist painted in pointillism.

Paintings capture moods, and the WMAP portrait is no exception. It shows the cooled-down form of a once-scalding universe, releasing long-pent-up energy into the gaps between atoms. The atoms were slightly clumped together, in patterns that depended in part on the geometry of space. Their particular arrangements indicated that they were happily settled into a flat, expanding hyperplane—with omega exactly equal to one. Perhaps they were especially content because they recalled a far more explosive period earlier on that flattened their vistas. But now they could move away from each other at a gentler pace, awaiting the day their gravitational attraction would compel them to reunite into myriad celestial bodies.

At a 2003 conference of the American Physical Society, physicist Michael Turner reveled in the high precision of the new data. He emphasized that, for the first time in the history of cosmology, researchers were able to perform exact-enough statistics to present their results with error bars (precise ranges of values). Turner also pointed out that the WMAP results ruled out the simplest inflationary models. He counseled, however, that there were other possibilities. "Fortunately, Andrei [Linde] had another 300 models left," Turner joked.

The combined power of the supernova and microwave background observations enables cosmologists to define a "concordance model" of the universe. Any theory that satisfies known results about the geometry, age, and content of the observable universe falls into this category. You would think that this would narrow things considerably. However, it still leaves the door open to diverse explanations—an inflationary era being only one of many possibilities.

Although we are now reasonably sure that the universe is flat, we still don't know exactly what *caused* it to be flat. (To recall, we mean here that the ordinary three-dimensional part of a four-dimensional Friedmann model is flat.) Was it born that way, molded through inflation, or smoothed out through another mechanism? Data from WMAP and other sources have converged on an estimate of 13.7 billion years for the *observable* universe, since the time of the Big Bang. But what about eras that may have preceded that colossal burst of energy? Perhaps there was no Big Bang singularity at all, just a transition between different phases of the universe. And could the observable universe be part of a greater whole, conceivably in higher dimensions? What of the dark energy that constitutes some 73 percent of the substance of the cosmos? Could it be a sign of something missing in our concept of gravitation? Could fundamental constants, such as Newton's gravitational constant or the speed of light, actually change over time? We will consider these disparate possibilities in chapters to come.

Finally, let's remember that, in addition to visible matter and dark energy, the cosmos appears to contain a third major component—dark matter. Readings from WMAP indicate that this hidden material represents 23 percent of the universe. Although theories abound, no one has yet developed a satisfactory explanation of what dark matter actually is. This enigma has grown even deeper with the recent discovery of an *entire galaxy* as inscrutable as the Cheshire cat.

Heinrich Wilhelm Olbers (1758–1840), originator of what is now known as "Olbers' paradox," the problem of why the night sky is so dark given that the universe is full of luminous sources. (Courtesy of the Bakos Observatory collection, University of Waterloo.)

Albert Einstein (1879–1955) was undoubtedly the greatest physicist of the 20th century. In developing the special and general theories of relativity, he updated Newton's ideas with more comprehensive descriptions of motion and gravitation. He spent his final years trying to develop a unified theory of all natural forces. (Courtesy of the AIP Emilio Segre Visual Archives, W. F. Meggers Collection.)

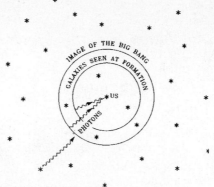

As shown here, the maximum distances traveled by incoming photons carve out spherical imaginary surfaces centered on Earth, distinguished by their time lags. If our telescopes were powerful enough, we could see an image of the universe shortly after the Big Bang itself. Beyond that shell, for an unbounded universe, photons would not have had enough time to reach us. Hence, the light we see in the sky is the sum of a *finite* set of sources, leading to darkness at night. (Illustration designed by Paul Wesson.)

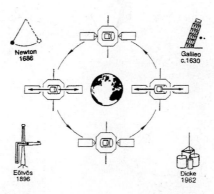

Various attempts to measure the equality of inertial mass and gravitational mass. Displayed counterclockwise are depictions of Galileo's legendary dropping of two different weights from the Leaning Tower of Pisa, an experiment by Newton involving a simple pendulum and the torsion balances of Eötvös and Dicke. Shown in the center is the design concept for STEP (Satellite Test of the Equivalence Principle), projected for launch into Earth orbit after 2011. Astronomers expect that this spacecraft will be able to measure the equivalence of mass within one part in 10^{18}. (Adapted from and by courtesy of NASA.)

The LIGO (Laser Interferometer Gravitational-Wave Observatory) detector in Hanford, Louisiana, stands guard for gravitational waves reaching Earth, resulting perhaps from cosmic cataclysms. Its two long arms ensure full spatial coverage of incoming signals. A second detector, located in the state of Washington, serves to confirm any disturbances measured by the first (and vice versa). (Courtesy of NASA.)

The 250-foot-diameter Lovell Telescope, at Jodrell Bank Observatory in Cheshire, England, is one of the largest radio dishes in the world. Recently, astronomers used the delicate instrument to detect the first-known dark galaxy, VIRGOHI21. (Photograph by Craig Strong, courtesy of the University of Manchester.)

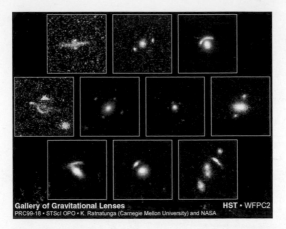

Assorted examples of gravitational lenses: situations in which the gravitational influence of closer objects (such as galaxies) distorts the light from distant bodies (such as quasars). Such distortion is a direct result of the warping of space-time predicted by Einstein's general theory of relativity. (Courtesy of NASA.)

The distribution of luminous and dark material (illustrated as a faint haze) in the cluster of galaxies CL0025+1654, about 4.5 billion light-years away. Employing the Hubble Space Telescope, astronomers used the gravitational lensing of more distant objects to determine the dark matter's layout. They found that the distribution of invisible material closely follows that of the visible galaxies in the cluster. (Image by J-P. Kneib et al., Observatoire Midi-Pyrenees and Caltech, courtesy of the ESA and NASA.)

The detector at the Sudbury Neutrino Observatory (SNO) lies more than 2,000 meters (one and one-quarter miles) under Earth's surface, where it awaits the rare impact of neutrinos from space. Data from this device have been used to ascertain the relative masses of various types of neutrinos and to help assess their relative contribution to the dark matter. (Courtesy of Ernest Orlando, Lawrence Berkeley National Laboratory and the Sudbury Neutrino Observatory Institute.)

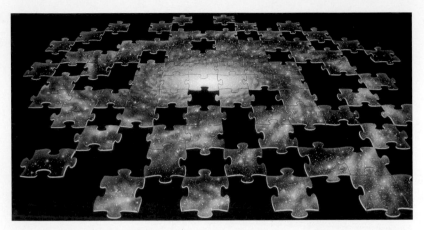

Like this shattered puzzle ("Galaxy Puzzle," illustrated by Lynette Cook), could the universe be fated to disintegrate? According to the controversial "Big Rip," scenario, billions of years from now the dark energy of the cosmos will tear its fabric apart. Everything in existence would be decimated—from mammoth galaxies down to tiny atoms. Such a catastrophic ending is but one of many conceivable fates for the universe. Theorists have suggested other possibilities—including a "Big Crunch," in which the cosmos someday collapses back to a singularity. (Copyright 1996 Lynette Cook, *http://extrasolar.spaceart.org*, used by permission.)

Cambridge University's Centre for Mathematical Sciences, with its playful, futuristic architecture, is a haven for scientists contemplating the origin, fate, and structure of the cosmos. (Photograph by Paul Halpern.)

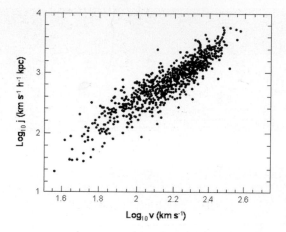

The Tully-Fisher relationship between the angular momentum per unit mass (j) and the spin velocity (v) of a typical spiral galaxy. Plotted on a logarithmic (base 10) scale to encompass the wide range of values, it indicates a slope that lies near one. This suggests a commonality between the ways that spiral galaxies acquired their spins. (Based on data from various sources as interpreted by Paul Wesson.)

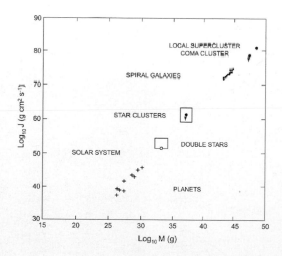

The angular momentum (J) and mass (M) of a typical astronomical object are correlated. To cover the large range in these parameters, from planets to the local supercluster, one can use a logarithmic scale as shown. The slope of the correlation depicted here lies between 1.7 and 2.0 and implies that astronomical objects acquired their spins in some similar fashion. (Based on data from various sources as interpreted by Paul Wesson.)

Paul Adrien Maurice Dirac (1902–1984) was one of the principal developers of quantum physics. His speculative Large Number Hypothesis, relating various parameters in the universe, led to the curious idea that gravity has changed its strength over time. (Photo by A. Bortzells Tryckeri, courtesy of the AIP Emilio Segre Visual Archives.)

Arthur Eddington (1882–1944), one of the foremost British astronomers, was among the first to understand the implications of general relativity. He organized solar eclipse expeditions to test the theory and contributed greatly to its popularization. In his later years he stirred up controversy with his ideas about the role of human thought in the shaping of physical concepts. (Courtesy of the AIP Emilio Segre Visual Archives, gift of S. Chandrasekhar.)

4   Darkness Apparent: The Hidden Stuff of the Cosmos

The bright suns I see and the dark suns I cannot see are in their place.

Walt Whitman (Song of Myself)

"I wish you wouldn't keep appearing and vanishing so suddenly: you make one quite giddy!" said Alice.
"All right," said the Cat; and this time vanished quite slowly, beginning with the end of the tail and ending with the grin, which remained some time after the rest of it had gone.

Lewis Carroll (Alice's Adventures in Wonderland)

THE CHESHIRE GALAXY

In Cheshire, cats tend to fade in and out of view—at least according to Lewis Carroll. Recently, astronomers working in that English county discovered that galaxies can similarly hide from sight. In the case of VIRGOHI21, an invisible galaxy detected at the Jodrell Bank Observatory in Macclesfield, Cheshire, only a grin of radio waves reveals its stealthy presence.

The shadowy creature first flashed its smile in a 2004 radio survey of the Virgo cluster. A team led by Cardiff University researchers Robert Minchin and Jonathan Davies found a rotating disk of hydrogen atoms approximately 100 million times the mass of the Sun. When they gauged the rotational speed of the pancake-shaped entity, it presented itself as if it were actually 1,000 times heftier—that is, 100 *billion* times the Sun's mass. Consequently, the astronomers concluded that more than 99.9 percent of the body is composed of unseen material. Observations shortly thereafter at the Arecibo radio observatory in Puerto Rico confirmed the Jodrell Bank picture. A close examination of the dark colossus found absolutely no trace of stellar objects. Hence, VIRGOHI21 is the first-known completely starless galaxy.

The Cardiff team has speculated about the invisible galaxy's origins. One possibility the team investigated was that it consists of material wrested from other galaxies in a cosmic tug-of-war. However, no nearby galaxies stood in the proper positions for exerting such tidal forces. "If it is tidal debris," Minchin and his colleagues concluded, "then the putative parents have vanished."

Another possibility the team investigated is that VIRGOHI21 is a gravitationally bound system whose hydrogen is too dispersed to clump into stars. Its scattered pockets of hydrogen may lack the critical density to fuse together and burn. Given the data the researchers found, this seemed to them the most likely option. Using their discovery as a model, they launched a concerted effort to find other invisible galaxies in space, for although VIRGOHI21 is the first-known completely dark galaxy, other galaxies brim with unseen material. This matter forms dark halos around the shining stars, protruding far beyond the visible bounds of these objects. The nature of this dark substance is largely unknown—a long-standing astronomical mystery.

Astronomers have suspected for many decades that the visible content of the universe falls far short of the amount that is apparently exerting gravitational force. Hints of this missing mass conundrum date as far back as the early 1930s, when (as we discussed) astronomers Jan Oort and Fritz Zwicky found unexplained behavior in galaxies, and clusters respectively. These ideas sat in the literature for years, however, until the pioneering work of Carnegie Institute astronomer Vera Rubin catapulted the issue to prominence.

THE MISSING MASS

Vera Rubin seems an unlikely revolutionary. The thick-spectacled, soft-spoken astronomer appears more the patient teacher than a radical firebrand. She has a methodical way of speaking that conveys her impressive attention to detail.

Her research advisor, George Gamow, was, in contrast, quite a showman. He loved to make bold claims and was not always so careful in his statements. He could be loud, funny, and boastful. A prolific amateur cartoonist, he relished teasing his colleagues through clever sketches and verse.

Despite the difference in styles, they shared an exceptional interest in pedagogy. She found him a brilliant lecturer with wonderful intuition, who even in his bravado often turned out to be prescient. This had a profound influence on her own career, which was geared toward education as well as discovery. (Perhaps there is no stronger statement of Rubin's ability to project enthusiasm for learning than the fact that all four of her children have Ph.D.s.) Indeed, Rubin came upon dark matter while developing an assignment for her students. Eight years after she received her doctorate from Georgetown under Gamow's tutelage, she was experimenting with creative ways of conveying information to her classes. As she recalled:

> In 1962, with my students at Georgetown University, we were looking at the [astronomical] literature. I was teaching at the graduate school. Most of my students worked at the Naval Observatory. I decided to have a class project in which we scanned stars beyond the solar system.

Rubin and her students plotted the velocities of stars in the outer reaches of the Milky Way versus distances from its central bulge. They fully expected velocity curves that would drop off with radial distance, like the foothills that surround a lofty mountain. Newton's laws, applied to concentrated systems, mandate such behavior. They proscribe slower speeds for the outermost orbits, where gravity tugs weakly. For example, distant Neptune revolves around the Sun at a far more languid pace than does Earth. By similar reasoning, the Georgetown class anticipated that remote stars in the Milky Way, far from its dense hub, would lag closer objects.

Wholly expecting velocity "foothills," the group was astounded when the velocity curves turned out to be "plateaus." They found no dropoff in speed, no matter how far from the galactic center they looked. Hence, the stars in the periphery of the galaxy moved much faster than they ought to, given the known matter distribution in the Milky Way. Some hidden substance seemed to exert an extra pull. As Rubin described this result, "The interesting thing about dark matter is that in order to have a flat rotation curve we have to have matter that we do not see, but [that is] spread farther out."

Rubin and the students wrote up their results in an article and submitted it to the prestigious *Astronomical Journal.* Shortly thereafter, Rubin received a disturbing call from the journal's editor. He was adamant that he wouldn't publish a paper with the names of students on it, which would break the journal's strict policy. Rubin insisted on keeping the names. The paper was published anyway, setting a marvelous precedent.

It took virtually another decade for the astronomical community to recognize the magnitude of the problem. By then numerous studies by Rubin and others had demonstrated that other galaxies similarly possess flat rotation curves; over vast distances from their centers, the velocities of their stars do not taper off but rather remain fairly level. Astronomers came to realize that something was drastically wrong. From one galaxy to another, and even in the spaces between, a huge portion of the cosmic bulk was simply invisible.

Rubin encountered controversy about whether or not her discoveries applied to less prominent galaxies. "Many astronomers said to me, 'If you look at faint galaxies, there's no dark matter,' but actually the opposite is true."

The difference between VIRGOHI21 and the galaxies examined by Rubin is that, while the latter have dark halos comprising approximately 90 percent of galactic mass, the former appears wholly dark. Nevertheless, masked operators seem to drive all these cosmic carousels. What *are* their secret identities?

Gravity's Lens

One prominent category of dark-matter candidates has a name befitting its supposedly Herculean strength. Called MACHOs (Massive Astronomical Compact Halo Objects), it consists of bodies (in the peripheries of galaxies) thought to be gravitationally powerful *en masse* but each too lightweight to shine. Like an army of tiny ants dragging fruit off a picnic table, these diminutive orbs would tug vigorously on visible stars. Conceivable MACHO types include brown dwarfs (nonshining stars), red dwarfs (very dim stars), white dwarfs, large planets, neutron stars, and black holes—in short, anything lacking a supply of hydrogen sufficiently concentrated to light its nuclear furnace.

In 1986, Princeton astrophysicist Bohdan Paczynski proposed a clever way of searching for and classifying potential MACHOs. Using the Einsteinian result that gravity bends light, he developed the technique of gravitational microlensing. Ordinary gravitational lensing occurs when an extremely massive object, such as a galaxy, stands directly in the path of emissions from a more distant body, such as a quasar. Rays from the latter bend around either side of the former, resulting in a double image (and sometimes even a multiple image). Astronomers literally see "twins" of the remote system they're observing.

If the lensed object and the lens lie directly on the same line of sight, then instead of multiple copies, a circular image appears called an "Einstein ring." The radius of the ring depends on the gravitational distortion of the matter bending the light. Therefore, it provides a means of gauging how much mass lies in the light rays' path.

Pacyznski adapted this method to account for minute concentrations of mass, showing that a customized form of gravitational lensing would be a suitable way of discerning small, otherwise invisible, bodies in galactic halos or elsewhere. The microlensing effect would be much more subtle than large-scale lensing—involving a slight brightening and dimming as the MACHO passed by—but still potentially discernible by computer-aided instruments.

Using gravitational microlensing to identify intervening MACHOs is akin to employing an eye chart to find the best-fitting glasses. In optometry, if an image looks distorted, the cause of such blurriness can be inferred and then corrected. Similarly, in astronomy any change in the appearance of a star could indicate the fleeting distortion caused by an intervening object. Working backward from the image, we can deduce the properties of the unseen agent, particularly its mass and size.

Armed with this powerful technique, several teams of astronomical detectives set out to sleuth for MACHOs in the early 1990s. One collaboration, headed by Charles Alcock and simply called the "MACHO group," conducted an extensive scan of millions of stars in the Large Magellanic Cloud (LMC)—a small satellite galaxy of the Milky Way. Another team, called EROS (*Expérience de Recherche d'Objets Sombres*—French for "The Experimental Search for Dark Objects") focused on both the LMC and the Small Magellanic Cloud—scanning a similarly vast array of stars. Each group looked for evidence of brightness variations, caused by invisible objects passing between the stars and Earth.

The immense numbers of stars needed for these surveys derived from the tremendously low odds that any one of them would be lensed by a MACHO. Both the star and the MACHO would need to line up almost perfectly (within one milliarcsecond of angle, to be specific)—an extremely rare coincidence. Observing millions of stars offered the teams much higher chances for success. Also, the greater the number of observations, the easier it was for the researchers to eliminate false calls. Many stellar bodies *naturally* vary in brightness—for example, the Cepheid variable stars employed by astronomers to gauge distances in the universe. The scientists needed to identity all these events (according to their characteristic profiles) and exclude them from their data. Only then could they be reasonably certain they were witnessing true MACHO microlensing.

In 1993 the teams were excited to report the first candidate objects. The solution to the dark-matter problem, at least for galaxies, began to seem close at hand. As the decade wore on, however, a glaring discrepancy opened up between the two groups' reports. The MACHO collaboration collected a bevy of positive results, pointing to an abundance of halo objects in the LMC. The typical mass of these bodies, about half that of the Sun, suggested they were stellar dwarfs (of some sort) radiating too weakly to detect. EROS, at first, found very few such objects. Thus, while the MACHO group's statistics portrayed galactic dark matter as consisting mainly of dim halo objects of half-solar size, the EROS team explicitly ruled out that possibility. Gradually, though, the MACHO collaboration began to downsize its estimates—drawing closer to what EROS found. Today, astronomers believe that only about a fifth of galactic dark matter is comprised of such dim stellar bodies—the composition of the remainder is still unknown.

What's more, identifying the missing material in galaxies is only a small facet of a much greater mystery. Unseen gravitational influences occur on all known astronomical scales. Therefore, it is clear that considerable quantities of dark matter lie beyond galaxies—in the spaces between them in clusters and in the voids between clusters themselves. Furthermore, evidence indicates that most of the invisible substance is *non-baryonic* in nature. Baryons are the stuff of atomic nuclei—protons, neutrons, and such. If the major part of the unseen material excludes atoms, it can't possibly be anything like stars. It must be something much more slippery to have thus far escaped detection.

CAPTURING THE NEUTRINO

Scientists often classify dark matter into two distinct categories based on its thermal properties: hot and cold. Hot dark matter involves fast-moving particles, such as neutrinos. Individually, these particles have negligible mass. They are so abundant, however, that even with small nonzero masses they'd collectively produce a significant gravitational effect.

Cold dark matter, in contrast, consists of slower-moving materials. MACHOs are often placed in this category. Non-baryonic examples include various classes of hypothetical particles, called axions and WIMPs (weakly interacting massive particles). The latter is a play on words—contrasting these diminutive constituents with "manly," stellar-sized MACHOs. Axions, WIMPS, and neutrinos are thought to be largely unseen because they interact so rarely with ordinary matter. Neutrinos, for instance, pass straight through Earth all the time. They are extraordinarily common particles, but because they are lightweight, electrically neutral, and impervious to the strong nuclear force, they have few opportunities for contact with other matter. For the most part they are oblivious to their surroundings, like someone in a sensory deprivation chamber. Their main mode of interaction lies in the weak nuclear force—through, for instance, the process of beta decay.

Identified but not fully understood in the late 19th century, beta decay is the name given to a process wherein neutrons break down into protons, electrons, and neutrinos. (Until the time of Pauli, physicists didn't know about neutrinos; he inferred their existence through conservation principles.) It is a common process—occurring, for example, during the transformation of radioactive isotopes.

The main source of neutrinos passing through Earth, though, is not radioactivity but rather the Sun, through its cycles of fusion. Inside the Sun's churning cauldron, the wispy particles are produced in great abundance. Passing though the Sun's outer layers and released into space, they rain down continuously on Earth and the other planets.

Originally, physicists thought these particles were completely massless. Special relativity implied, therefore, that like massless photons they moved at the speed of light. Furthermore, they were thought to consist of only a single variety—as uniform as the desert sands; but puzzling results from neutrino detectors capturing emissions from the Sun would eventually challenge these assumptions.

In 1967, Raymond Davis of the University of Pennsylvania inaugurated the first observatory designed for collecting solar neutrinos. The apparatus was very simple—a mammoth tank filled to the brim with 100,000 gallons of chlorine-based cleaning fluid. As endless droves of neutrinos swarmed through the tank, each had a slim chance of colliding with one of the chlorine atoms and transforming it into radioactive argon. The new isotope would then announce itself through a characteristic emission. Based on the average spacing between atoms, the size and geometry of the tank, Earth's distance from the Sun, the dynamics of solar processes, and other factors, Davis and his assistants could then compare the expected influx of neutrinos with the actual number of events.

If the tank had been exposed to the atmosphere, the experiment would have been hopeless. Cosmic rays of all sorts bombard our planet all the time. It would be impossible to know which collisions were the "real deal" and which were phonies. Some kind of "bouncer" was required to keep the hoi polloi out of the club and let in only well-pedigreed neutrinos. Fortunately, the experimenters found a natural solution. They realized that, if they placed the apparatus deep underground, Earth's mighty strata of rocks and soil would keep a firm guard against unwanted intruders. Only neutrinos would have the slinky skills required to penetrate the rocky layers. Thus, the team lowered the tank into the everlasting darkness of the Homestake gold mine, almost a mile beneath the Black Hills of South Dakota.

After the experiment was running for some time, Davis began to notice a severe discrepancy between the expected and actual numbers of neutrinos. Something like two-thirds of the anticipated crowd simply wasn't showing up. There was either a fatal flaw in the apparatus or a gross misunderstanding of how solar dynamics worked. When subsequent experiments at other subterranean detectors around the world confirmed the deficit found at Homestake, researchers began to rethink their suppositions.

Theorists dusted off certain alternative neutrino theories that imagined them as more versatile characters. Could the neutrinos produced along with electrons in beta decay represent only one of several different types? Might there be a special kind associated with muons (particles similar to electrons but considerably heavier) as well? Later, with the discovery of tauons (even heavier than muons), this scheme was augmented to include three different types: electron neutrinos, muon neutrinos, and tau neutrinos. Perhaps under particularly intense conditions, such as those of the Sun's nuclear furnace, neutrinos "oscillated," or changed from one breed into another. If that were the case, it would explain why the bulk of solar neutrinos could not be detected—they were simply wearing different guises. Furthermore, neutrino oscillation models imply that they must possess different masses. In transforming, they shift from one rung of a ladder of masses to another. Thus, they cannot be absolutely massless.

This realization, combined with a fervent interest in resolving the dark-matter quandary, set off a race to pin down the masses of the neutrino varieties. Recognizing that these values would be extremely small, researchers hoped that delicate statistical measures could help distinguish them from zero. Three neutrinos, with various degrees of heft, would offer a handy trio of dark-matter components. If the combined weights of electron neutrinos turned out to be insufficient to address the missing-matter dilemma, could muon and tauon neutrinos offer enough bulk to do the job?

These burning questions motivated the construction of the largest neutrino detector to date, the Sudbury Neutrino Observatory (SNO) in Ontario. A converted nickel mine, more than a mile and a quarter beneath the Canadian soil, houses a gargantuan acrylic vessel filled with 1,000 tons of ultrapure heavy water (with deuterium instead of hydrogen) surrounded, in turn, by a reservoir of ordinary water. Thousands of photomultiplier tubes (high-precision light sensors), arranged like sentries around the tank, stand guard for the unique flashes of neutrino collisions. Each type of neutrino, as it slams into a deuterium atom, produces a characteristic signature. These signals are collected and statistically analyzed, offering a sample of the Sun's varied output.

In 2001 the SNO collaboration—a team of Canadian, American, and British scientists headed by Art McDonald of Queen's University—announced the first results. In a stunning breakthrough, they found enough events to resolve the solar neutrino problem and prove that these particles come in three varieties. This finding, along with results by the Liquid Scintillator Neutrino Detector experiment at Los Alamos, helped establish the mass differences between each of the types.

Based on these and other critical results from around the world, today scientists believe that the neutrino trio constitutes a segment, but not a major component, of the dark matter in the universe. Even tallying neutrinos along with MACHOS yields far too little mass to fill the gap. Attention has shifted to some of the other candidates—particularly axions and WIMPs.

MASKED MARAUDERS

Like the Mixmaster universe, axions are whimsically named after a commercial product—in this case a brand of laundry detergent. Particle physicist Frank Wilczek couldn't resist the opportunity to draw a chuckle by using that name. Fittingly, they are believed to exist because they cleansed the strong nuclear interaction of one of its original puzzling properties. To understand what property was removed, let's revisit the topic of symmetry breaking in the early universe.

One type of symmetry in nature is called CP (charge-parity) invariance. That means that, if the charges of an interacting system are reversed (from positive to negative, or vice versa) and the system is reflected in a mirror, it would look exactly the same as the original. For a single rotating particle, the latter process—known as parity reversal—involves switching the direction of rotation (from clockwise to counterclockwise, or the converse). Thus, in short, by flipping pluses to minuses and clockwise spins to counterclockwise spins, the system under consideration should return to its initial state.

The highly successful theory of the electroweak interaction includes a term that explicitly *violates* CP invariance. The term entered the theory because of experiments showing that CP violation is an inherent feature of the weak interaction. For certain processes involving the weak interaction, changing the signs of all the electrically charged particles and then reflecting them in the mirror leads to a system as different from the original as right-handed gloves from left-handed gloves. This disparity can be seen, for example, in the decay of certain elementary particles called kaons.

Theorists hoping to extend the standard model to describe the strong interaction found the need to include a similar term. Yet they ran up against a brick wall. Strong processes all respect CP symmetry. Like Isaac Asimov's perfectly programmed robots, under no circumstance will they break this rule.

A good example of this property involves the neutron. Strong-force models predict that this particle should possess a physical feature called an electric dipole moment. According to electromagnetic theory, the actions of electric and magnetic fields can be grouped into "moments": dipole, quadrupole, and so on. The electric dipole moment constitutes the effect of pairs of equal but opposite charge—separated, like barbells, by a certain distance. The vector representing this moment points in the direction of the positive end of each pair. For magnetism—where the basic unit is poles—the magnetic dipole moment is the analogous quantity, pointing toward the north pole of each magnet.

Scientists have long known that neutrons possess *magnetic* dipole moments. Theoretically, they might be expected to have *electric* dipole moments as well. Modeling neutrons as sets of charged pointlike quarks would suggest such electric field components. If neutrons did have electric dipole moments, though, they'd violate CP symmetry, because by inverting the neutron's internal charges and flipping its rotation, the neutron's electric dipole moment would no longer point in the same direction as its magnetic dipole moment. Therefore, altering both charge and parity would effectively produce a different particle.

However, as firmly established by repeated experimentation, neutrons have no electric dipole moment at all. Consequently, they—along with any system bound by the strong force—emphatically don't violate CP. This inescapable issue, called the "strong CP problem," led many theorists to wonder if the strong force's propensity for CP violation could have been stolen somewhere along the way.

Which of nature's agents, then, served as the thief? Roberto Peccei of UCLA and Helen Quinn of Stanford have identified the culprit as the axion. According to Peccei-Quinn theory, at a certain point in the universe's history the strong interaction experienced a spontaneous symmetry breaking that transformed it from a CP-violating to a CP-preserving theory. The axion constitutes the particle created by this process—the marauder that wrested away the original property.

If axions exist, they'd be extremely common but tremendously hard to detect. Like neutrinos, they'd each have an exceptionally low mass (perhaps one-trillionth that of the electron), would have no charge, and would interact only rarely. In contrast to neutrinos, though, they'd move slowly enough to play a role in structure formation. For these reasons, many cosmologists find them prime suspects for a large segment of space's unseen material—because they'd be so plentiful, their minuscule masses would add up to a heaping helping of dark matter.

In recent years, physicists have launched a number of experiments to search for these stealthy particles. Some of these experiments, such as those at Brookhaven National Laboratories (New York), CERN, and the University of Tokyo, have focused on finding axions amid the intense radiation of the Sun. The postulated mechanism for such production of axions is a process called the "Primakoff effect," entailing the bouncing of energetic photons off electric charges. Pierre Sikivie of the University of Florida suggested a means of detecting axions by reversing this procedure: using powerful magnetic fields to convert axions back into photons. Although this method seems promising, no such solar axions have been found so far.

Other searches have focused on the possibility of detecting axions forged in the flames of the primordial cosmos. To that end a promising endeavor has been launched at the Lawrence Livermore National Laboratory (LLNL) in California. Headed by physicists Leslie Rosenberg and Karl van Bibber, a team from LLNL, MIT, UC Berkeley, the University of Florida, and the National Radio Astronomy Observatory are preparing the most extensive sweep for axions ever attempted.

Using Sikivie's technique, the team has designed a high-precision apparatus involving a sensitive radio receiver attached to a tunable microwave cavity, surrounded by a superconducting magnet. The idea behind the experiment is straightforward. The strong magnetic field would transform rogue axions into energetic photons. Then, as the cavity's frequency is slowly adjusted, it would resonate at some point with the photon's frequency, which, in turn, would produce a weak signal that would be amplified with a special quantum device until it was detectable. Just to give you an idea of how faint the signal would be—it would be more than 100,000 times weaker than the broadcasts from the *Pioneer 10* space probe as it reached the edge of the solar system!

Rosenberg is optimistic that this special apparatus, upgraded from earlier detection devices, will prove the ultimate arbiter for whether or not axions form the lion's share of dark matter. Even if it turns out that axions constitute but a minute component, he believes his group's detector will reveal their presence and gauge their relative influence.

"The new upgrade will be sensitive enough to detect even the weakest signals," says Rosenberg. "In other words, this upgraded search will likely be the definitive search . . . and no matter what we discover, it will be illuminating. Either we will find the axion, proving that it exists and is part of the cosmological evolution of the universe, or we won't. . . . If there is no axion, there must be entirely new physics—some strange, new physics that we cannot as yet fathom. And that, too, would be very interesting indeed."

BATTLE OF THE WIMPS

With MACHOs and neutrinos looking increasingly unsuitable to be major dark-matter components and axions still just a conjecture, some physicists have climbed even farther out on the limb of speculation and made a case for WIMPs. This is a broad category, covering a host of theoretical entities that share an intense dislike for interacting with ordinary matter. One major advantage of WIMPs over neutrinos and axions is that they'd be much heavier. So they could be somewhat less common but still massive enough to make a profound difference.

Typically, experimental searches for WIMPs have been conducted in places comparable to neutrino-hunting expeditions—deep underground. The reasons are similar. In both cases, ground-level searches would be inundated with a flood of extraneous particles from space. Countless superfluous events would need to be screened out. Therefore, subterranean locales, such as caverns and mines, offer greater protection against cases of mistaken identity and improved chances for genuine results. For example, one leading search for WIMPs is based under the Oroville Dam (California)—the tallest in the United States. Another is set more than two-thirds of a mile underground in the Boulby salt mine (North Yorkshire, England)— the deepest in Europe.

Even in the perpetual darkness, however, unwanted particles can intrude, so researchers try to choose locales far away from known veins of radioactive ore. As an added precaution, they often house their apparatus in thick layers of either lead or copper. Finally, some experimenters have dramatically reduced thermal noise by cooling their detectors close to absolute zero.

The main method for detecting WIMPs is called nuclear recoil, which involves rare collisions that cause atomic nuclei to shift slightly back and forth, giving off photons in the process. The minute quantities of energy released in such jolts can be detected in certain standard ways, including ionization and scintillation. The former method, used at Oroville, entails the release of outer electrons from atoms. Germanium, a hard, gray-white material, has been found particularly effective for this purpose. Scintillation, on the other hand, involves special types of material that absorb energy and release detectable flashes of light. The Boulby team, for example, has used two types of scintillating materials: sodium iodide and liquid xenon. Photomultipliers placed around these substances can pick up their characteristic flashes. Computers then analyze these signals to discern the possible signatures of WIMP collisions.

Currently, most researchers working in the field are still waiting for the first signs of WIMPs. However, a group from Italy, called DAMA (particle DArk MAtter searches with highly radiopure scintillators at Gran Sasso), has already claimed success. Led by Rita

Bernabei of the University of Rome, the team first announced positive preliminary findings in 1996. They have since updated their methods and have continued to publish affirmative results.

The DAMA technique focuses on the possibility of a WIMP wind from the galactic halo. Our planet's annual revolution around the Sun and the motion of the solar system through the Milky Way would make this wind periodic. Certain times of the year we'd be speeding into it, and other times we'd be heading away from it. Thus, it would manifest itself as a cyclic flux (amount per area) of weakly interacting particles pounding down on Earth at a variable rate. It is this fluctuating storm of WIMPs that the Italian group claims to have detected.

But are the DAMA detectives on the right track? Close on their tail has been another group, conducting the Cryogenic Dark Matter Search (CDMS) experiment, based originally in a tunnel under Stanford University and now running in the Soudan mine in Minnesota. That team has conducted a comparable analysis with its own detectors and has reached a diametrically opposite conclusion. Not only do the CDMS researchers see no evidence of an annual modulation, they have also conjectured that the particles found by DAMA could be ordinary neutrons, rather than exotic particles. Both groups are currently gearing up for the next stages of the battle, showing that WIMP research is not for wimps.

Regardless of whether or not WIMPs have actually been observed, theorists have lined up an ample selection of other candidates, including bloated massive neutrinos of an unknown sort, bosons beyond the current radar screen of detection, and especially sparticles, the massive supersymmetric companions of ordinary particles.

SPARTICLES AND SHADOW WORLDS

Supersymmetry is a theoretical link between the two great kingdoms of particles: fermions and bosons. An example of the former is an electron and of the latter a photon. Basically, fermions comprise the constituents of ordinary matter, while bosons relate to the forces between them. The main difference between the two types is *spin*. In this context, spin relates to a particle's set of possible transformations in an abstract space. Some particles, such as the Higgs, are always in the same spin state, called spin zero. Others, such as electrons, have two possible spin states, called up and down (also known as +1/2 and −1/2). Yet others have three (−1, 0, +1) or four (−3/2, −1/2, +1/2, +3/2) possibilities. Fermions are defined as those particles with half-integer spins—granting them an even number of possible spin states. Bosons, in contrast, possess integer spins—generally allowing them an odd number of possible states.

In traditional field theory, no matter how hard you tried, you couldn't transform a half-integer spin state into an integer state. Quantum physicists accommodated to this principle long ago and developed separate statistical methods for each category. This seemed satisfactory until the advent of string theory.

String theory arose in the early 1970s as a description of the strong force but has since been framed as a road to unifying all natural interactions. In its original incarnation, it described bosons as flexible "rubber bands" of energy and fermions as their end points. This description was intended to represent the property of the strong force to restrain nuclear particles if they move too far away from each other, but it allows them abundant freedom if they are relatively close. Mathematical analyses of bosonic strings demonstrated how their vibrations could represent various energy levels. This is similar to how various vibrations of stringed instruments produce distinct pitches. If music has harmonics, why not the fundamental constituents of nature?

Given the neat mathematics of bosonic strings, physicists began to wonder if fermions could be described similarly. The problem was how to transform an integer spin theory into a half-integer spin theory. Pierre Ramond, of the University of Florida, put forth an intriguing solution. He postulated a hitherto-unseen symmetry of nature, encompassing both fermions and bosons and allowing for half-integer spin transformations from one to the other. Jean-Loup Gervais and Bunji Sakita of the City College of New York independently proposed a related version around the same time, which was further developed by Julius Wess of the University of Munich and Bruno Zumino of CERN. Because it superseded known symmetries, this clever transformational scheme became known as supersymmetry.

Soon, supersymmetric string theory, or "superstrings" for short, acquired acclaim in some circles and notoriety in others, when physicists John Schwarz and Joël Scherk showed that it naturally predicted a spin-two particle with particular properties. If you identify this particle as the carrier of the gravitational force, the resulting theory yields the equivalent of general relativity. Proposing superstrings as a candidate "theory of everything," Schwarz speculated that it could unite other interactions with gravity into a single model.

Meanwhile, Wess, Zumino, and other theorists applied supersymmetry to standard particle physics (without strings) to produce a generalization of general relativity called supergravity. Supergravity's star rose in the late 1970s and early 1980s until researchers in the field encountered a host of formidable mathematical difficulties. Then in 1984, Schwarz, along with Michael Green of the University of London, demonstrated in an influential paper that superstring theory was free of many of these ailments, bolstering its status. Encouraged by Schwarz and Green's results, a number of prominent physicists joined the superstring bandwagon. With its alluring mathematical properties, superstring theory seemed to offer a whiff of the physics of the future. Leading physicist Ed Witten proclaimed that "string theory is a piece of 21st century physics that happened to fall into the 20th century."

Nevertheless, many traditional physicists expressed strong doubts. Dismayed by the poor prospect for experimental confirmation of the theory any time in the conceivable future, they cringed at the thought of losing a generation of theorists to such a speculative approach. As journalist John Horgan said:

> Unfortunately, the microrealm that superstrings allegedly inhabit is completely inaccessible to human experimenters. A superstring is supposedly as small in comparison to a proton as a proton is in comparison to the solar system. Probing this realm directly would require an accelerator 1,000 light-years around. This problem led the Nobel laureate Sheldon Glashow of Harvard University to compare superstring theorists to medieval theologians.

A continuing subject of controversy concerns the strange doppelgangers predicted by any supersymmetric theory—not just strings. Applied to standard elementary particles, supersymmetric methods yield counterparts with opposite spin properties. Physicists denote these hypothetical companions by tacking on an initial "s" or a final "ino"—referring, respectively, to the bosonic equivalents of fermions or the fermionic mates of bosons. Hence, a supersymmetric twist turns quarks and electrons into squarks and selectrons and transforms photons and W particles into photinos and winos. The supersymmetric companions of particles are in general known as sparticles.

None of these particle counterparts have thus far been found in nature—which some see as a failing and others as an opportunity. Does the lack of sightings signify that supersymmetry is based on a false premise? Or does it simply mean that we haven't yet searched at high enough energies? If the latter is true, supporters argue, sparticles would be extremely massive and could well constitute the missing ingredient of the universe.

Foremost among the supersymmetric WIMP candidates is a peculiar hybrid called the neutralino. Rather than the partner of a single particle, it is the supersymmetric soulmate of several different bosons. (Apparently polygamy is allowed in the particle world.) It forms an amalgamated companion of the photon (photino), Z boson (zino), and Higgs boson (higgsino), mixed together in a quantum state.

The lightest version of the neutralino fits the profile of a dark-matter candidate well. First, unlike numerous short-lived particles, it is believed to be stable. Second, it is thought to interact only through the weak and gravitational interactions. Finally, its mass is in the appropriate range. For these reasons, many physicists have tagged it as a likely bet. In coming years we'll see if they win their wagers.

Though WIMPs are bizarre, the minds of theorists have produced even stranger possibilities. Consider the case of shadow matter: particles that respond only to gravity. They'd be invisible not just to optical telescopes but to all light-sensing instruments. Participating in no known decays or interactions, they'd be imperceptible to standard detectors. Neither chemical transformations nor nuclear recoils would herald their passage through Earth. Only extremely sensitive gravity wave detectors, well beyond current capability, would stand a chance of revealing these ghosts.

One hypothetical type of shadowy material, called "mirror matter," was proposed as a dark-matter candidate by physicists Rabindra Mohapatra and Vigdor Teplitz in 1999. Mirror matter would consist of particles of opposite chirality (handedness) than their counterparts in the conventional particle world—for example, right-handed "mirror neutrinos" corresponding to ordinary left-handed neutrinos. Symmetry principles would preclude electroweak interactions between mirror particles and garden variety materials; only gravity could provide a connection. Thus, a particle Alice could never step foot in the looking-glass world; she could only sense its gravitational pull.

Some researchers have speculated that somewhere in space— maybe even in the "Cheshire galaxy"—entire planets could be shaped from substances impervious to light. Well beyond our awareness, shadow aliens could be conducting their daily business, eating their shadow food, and basking in the unseen energy of their shadow stars. They would likewise be oblivious to our own type of matter—until, someday, pulsating gravitational signals link our disparate civilizations. Such an event would be one of the most outlandish resolutions of the dark-matter dilemma.

THE ESSENCE OF QUINTESSENCE

The discovery of cosmic acceleration has triggered a search for yet another type of missing substance—more precisely, an unknown source of energy. In many ways, dark energy is even more mysterious than dark matter. No material with which we're familiar exhibits an antigravitational force. We have discussed the repulsive properties of hypothetical objects with negative mass—but imagine a force with enough muscle to push all the mass in the universe apart!

In H. G. Wells's classic novella, *The First Men in the Moon*, he describes a substance called "Cavorite" that enables spacecraft to overcome gravity and effortlessly lift off from Earth. In the tale an inventor named Cavor "believed that he might be able to manufacture this possible substance opaque to gravitation out of a complicated alloy of metals and something new. . . . If one wanted to lift a weight, however enormous, one had only to get a sheet of this substance beneath it and one might lift it with a straw."

Could quintessence represent a kind of Cavorite that is able to counteract universal gravitation and accelerate the universe? Is it possible that this substance dominates certain phases of the universe but not others? Could its influence even be *growing* in strength?

Many physicists don't think of dark energy as a substance at all, in the traditional sense. Rather, they view it as a vacuum energy—the impact of the sea of virtual particles that pop in and out of the froth. From this perspective—rather than an independent, detectable quintessence—it is simply the lambda term, an essential quantum feature of space.

Indeed, long before the discoveries of Perlmutter, Schmidt, and their colleagues, the cosmological-constant issue was well familiar to field theorists. They came upon the problem from a different direction—their calculations produced far too high a value. Yet at the time most cosmologists assumed that lambda must be zero. How could these disparate values be reconciled? As Alan Chodos, associate executive officer of the American Physical Society, has remarked, "The old question was why is it zero? Now it is, why is it almost zero and incredibly tiny?"

This dilemma perplexed numerous researchers, such as the young Indian physicist Raman Sundrum, currently at Johns Hopkins. Sundrum carried out his Ph.D. studies of this issue under Lawrence Krauss and Mark Solvay at Case Western University well before the discovery of universal acceleration. As he recounted:

> At the time there were only bounds on the acceleration, nobody had actually seen acceleration. We saw the expansion, but not the acceleration. But these bounds were already a problem in the sense that the bounds said, whatever the acceleration was it was very small, whereas theory preferred very big. And so there was already a puzzle, that got a lot more interesting when we actually saw that there was not just a bound, but actually some finite acceleration.

Sundrum delved into this riddle with gusto, trying to find an explanation in the realm of field theory. Each particle model carried with it a gumbo of masses, interaction strengths, and other parameters. By stirring these ingredients, you could try to cook up the most savory stew—matching as much as possible the flavor of observed astronomical data. In particular, you might create the magic broth that yields a delectable value of lambda. As Sundrum realized, "The cosmological constant is incredibly sensitive to microscopic physics."

Standard field theories, however, generally serve up whopping plates of lambda, too large for general consumption. Finding these theories unpalatable, theorists sought a way of balancing out this excess with a factor that contained a *negative* cosmological constant. Such considerations led various researchers, including Sundrum and physicist Lisa Randall, to postulate the background geometry of the cosmos as a type of manifold called an "anti-de Sitter space." Such a space is warped (along a fifth dimension) rather than strictly flat. With deep connections to new versions of superstring theory, particularly the approach known as M-theory, the Randall-Sundrum model represents a popular new means of addressing the dark-energy dilemma.

Many field-theoretical approaches (including Randall-Sundrum) postulate that gravity takes a different form on various scales, which would naturally explain why gravity seems strictly attractive on the local level (the solar system, say) but harbors a repulsive component much farther out. Physicists like Eric Adelberger of the Eöt-Wash group have engaged in high-precision torsion balance testing of this hypothesis but have found no deviation so far from the standard law of gravity. Nevertheless, theorists have pressed on with a variety of alternative gravitational models.

The Big Rip

If antigravity turns out to be a dynamic property of the universe, one of the most frightening possibilities is it snowballing beyond control. The scenario unfolds like this. In the early universe, matter dominated dark energy. As the universe expanded, dark energy caught up and eventually slightly tipped the balance. But suppose this growth is far from over, releasing new reservoirs of repulsive force. Over time, gravity would increasingly cower before its towering competitor. Large structures such as clusters, then galaxies and smaller entities, would break apart. Ultimately, this would lead to the complete decimation of every shred of material in the universe—in other words, a "Big Rip."

The Big Rip joins the wailing chorus of other apocalyptic scenarios, including the "Big Crunch" and the "Big Whimper." Such dire circumstances—though in the far, far future—attract great interest. Like drivers slowing down to view a wreck or watchers of disaster movies, most people are curious about catastrophes. The demise of everything in existence certainly fits the bill.

Which will it be then—a splitting apart, a smashing together, or a more quiescent ending? Is being torn into pieces worse than being pulverized to a point? Clearly that is a matter of personal preference. Fortunately, we don't have to worry about this for many billions of years—long after the death of the Sun, when all life on Earth will presumably be extinguished.

One of the major differences between the various scenarios has to do with the prospect for communication with other galaxies before the ultimate cataclysm. What if there are alien civilizations beyond the Milky Way attempting to contact kindred beings? Suppose their beacons were somehow powerful enough to reach us, albeit in the far future. Could we ever hope to receive their signals?

The answer depends on the relative motion of galaxies in space. In the Big Crunch scenario, there would eventually be a limit to galactic recession. Billions of years from now, distant galaxies would stop traveling away from us and start to move closer. From that point on, they'd gradually become more prominent in the sky, facilitating the possibility of intergalactic communication. As the universe grew smaller and smaller, perhaps proximity (and necessity) would spur an intergalactic civilization. Conceivably, the greatest minds in such a culture would join together in an attempt to circumvent the complete destruction of the universe.

Combating universal catastrophe would be a less urgent matter in the Big Whimper picture, which represents a slow, continued expansion. Thus, perhaps, the need for intergalactic cooperation to address a common danger would be less immediate. That is fortunate, because as the universe continued to grow, its galaxies would be increasingly dispersed. The sky would be dotted with fewer and fewer spirals, ellipticals, and other ensembles of stars. Eventually all that could be seen, even with a telescope, would be the local group of galaxies; the others would lie beyond visual range. Consequently, any alien signals sent by remote extragalactic civilizations would never reach us. We'd be cosmic hermits, separated from other systems, until all the stars in the Milky Way burned out—turning into white dwarfs, neutron stars, and black holes. When doomsday finally arrived, it would be extraordinarily dark and lonely.

The road to a Big Rip would greatly exacerbate this isolation. The lambda force would pull the universe apart at breakneck speed, like a glutton attacking a bucket of chicken wings. Sooner than in the previous scenario, signals would be unable to span the increasingly formidable gaps. Hence, if we have any chance of intergalactic communication we'd best attempt it expeditiously; otherwise, we may someday wake up and find that it is too late.

If it seems ambitious to talk about possible events billions of years hence, you've perceived correctly. With the ease of meteorologists forecasting one or two days ahead, cosmologists feel comfortable projecting eons into the future. Yet if you read the fine print, many of the predictions are based on the proposition that the known laws of nature, as measured from Earth, must hold true for all places and all times.

But what if the principles of nature themselves evolve, like the stunning metamorphoses of "Darwin's finches" on the Galapagos Islands? Unless we perfectly understood these changes, we'd be in little position to make projections. Indeed, some of the exciting new cosmological theories posited to resolve the dark-matter and dark-energy riddles are based on the astonishing notion that nature's very "constants" could alter throughout the ages.

5   Ever-Changing Moods:
Did Nature's Constants Evolve?

The Sefer ha-Temeneh (13th century mystical text) teaches that in the seventh millennium there sets in a gradual and progressive retardation in the movement of stars and the spheres, so that the measurements of time change and become longer in geometrical progression. . . . Hence [mystics] arrived at truly astronomical figures for the total duration of the world.

Gershom Scholem (*Kabbalah*)

Whether or not it is clear to you,
No doubt the universe is unfolding as it should.

Max Ehrmann (*Desiderata*)

THE LEXICON OF NATURE

Physicists and astronomers like to frame natural laws in terms of simple mathematical equations. Labeling specific features of nature with symbols, they seek unambiguous relationships between such values, unalterable over time. Einstein's famous equation relating energy to mass and the speed of light squared constitutes an example of this tendency, and there are countless others. The hallmark of such equations is that they can be tested again and again, never failing to yield the same result. Therefore, like a warm bowl of porridge in the morning and a steaming cup of cocoa in the evening, they offer the comfort of regularity.

Within the framework of equations, scientists like to distinguish two types of elements: variables and constants. Traditionally, constants aren't supposed to alter over time. For instance, when Einstein composed his energy-mass relationship, he fully expected that the speed of light, a constant, would be the one permanent factor. So for any type of material under any kind of circumstance, he posited that this value would never change.

Another well-known equation is Newton's inverse-square law of gravity. It too contains a seemingly enduring fixture of nature, the gravitational constant. When Einstein proposed general relativity as a theory of greater scope than that of Newton, he kept the same constant. Both cases represent invariant relationships—first formulated theoretically and later confirmed by observation as being correct descriptions of our world.

However, there is another kind of law, logically distinct from the type to which Newton's and Einstein's theories belong. When we examine nature's vast array of phenomena, we sometimes observe patterns of a wholly different sort. Rather than the products of predictable equations, they constitute much subtler relationships that sometimes only the remarkable organizational capacities of our minds can perceive.

Consider, for example, the intricate designs of seashells and the elaborate lacework of snowflakes. Neither of these is governed by immutable equations. Instead, these spectacles emerge through self-organization—wonderful instances of order stemming from chaos. In the first case, the Fibonacci sequence of numbers, formed by adding each pair to produce the next (1, 1, 2, 3, 5, etc.), serves to characterize the length of successive turns in a spiral. In the second, the molecular geometry of water delimits the six-pronged symmetries of icy shapes. In each case, mathematical features manifest themselves in surprising ways.

Certain musical refrains seem more pleasing to our ears, certain artistic conceptions more beautiful to our eyes. We would recognize the first four notes of Beethoven's fifth as played on everything from an accordion to a tin whistle—instruments with radically different material compositions. Michelangelo's most famous statue could be carved out of ivory or soldered from sheets of steel; we'd still know the lad as David. No physical law allows us to recognize and enjoy such creations; rather our magnificent gray matter somehow does the job.

Sometimes the predilections of mathematics and the preferences of our senses even coincide. For example, the Greeks favored the Golden Ratio in art and architecture because they knew that this unique number produced pleasing relationships. Thus, the grandeur of the Parthenon attests to both the masterful structures wrought by mathematics and our unique abilities to perceive and make use of such constructs. We can't always rely, however, on our pension for *penchant* detecting patterns. In some cases it can prove misleading instead of enlightening. A gambler might notice that the first five spins of a roulette wheel land on the corresponding numbers of the Fibonacci sequence. He bets his fortune on the sixth, then glumly empties his pockets after his hypothesis proves wrong.

Einstein's equations of general relativity harbor many distinct solutions. Some of these represent real aspects of our universe. Others apparently constitute false leads. It is our pattern-discerning ability that helps us choose among these. But only the test of time—continued observation using increasingly powerful experimental tools—will corroborate our theories.

In the film *Pi*, the mentally prodigious protagonist begins to see patterns in *everything*—from biblical writings to the stock market. Convinced he has found a grand scheme that underlies all creation, he pushes ever further. All material comforts succumb to his frantic quest. Finally suffering a nervous breakdown, he decides to trade his numerical search for inner peace.

Whenever we perceive natural patterns, we must reasonably ask ourselves if they represent deep truths or are simply the products of our overactive imaginations. Sometimes our hunches guide us to extraordinary discoveries. Other times they lead us to dead ends. Still our curiosity drives us to their very limit, with the hope of arriving at ultimate knowledge.

DIRAC'S BOLD IDEA

Oxford mathematician Ioan James has detected a curious pattern among members of his own profession. Surveying their personalities and achievements, he has arrived at the conclusion that particular kinds of social deficits, possibly neurological in origin, often correlate with the focus needed for monumental discoveries. Perhaps certain personal, physical, or psychological limitations help concentrate the mind. Or, alternatively, perhaps development in some areas of the brain comes at the sacrifice of other areas. Thus, for instance, mathematical geniuses may not always make the best conversationalists.

The physicist Paul Adrien Maurice Dirac, one of the subjects of James's study, was a notoriously inscrutable 20th century thinker. He was a mystery even to many of his closest friends. "Nobody knew him very well," recalled physicist Engelbert Schucking, who encountered Dirac at various conferences. His colleagues joked that he rarely said anything more than "Yes," "No," or "I don't know." Legends swirled around him like tales spun about uncharted islands. They typically focused on his economy of words and his solemn dedication to pure science. In one such story, Dirac had just finished reading Dostoevsky's *Crime and Punishment*. Asked about his impressions of the classic Russian novel, he had only one comment: "It is nice, but in one of the chapters the author made a mistake. He describes the Sun rising twice on the same day."

Another time Dirac was delivering a lecture in his usual crisp and precise style. He never minced words and always planned each sentence meticulously. After the talk, there was a question-and-answer session. Someone raised his hand and said, "Professor Dirac, I do not understand how you derived the formula on the top left side of the blackboard."

"This is not a question," Dirac curtly responded. "It is a statement. Next question, please."

Dirac's rigidity in conversation contrasted with his extraordinary brilliance and creativity in discerning the properties of the universe. In the early days of quantum mechanics, his agile mathematical mind rapidly encompassed radical new ways of interpreting physics. He developed theories and ideas so fantastic, such as a negative energy sea that fills all of space, that they knocked the breath out of his colleagues. This concept relates to arguably his most important contribution to physics, the Dirac equation, proposed in 1928. The Dirac equation offered a quantum, relativistic description of an electron, encompassing properties such as its mass, charge, and spin. It predicted the existence of positively charged counterparts to electrons. Known as positrons, they were first experimentally detected in 1932. For his pivotal scientific contributions, Dirac was awarded a Nobel Prize the following year.

In 1937, Dirac applied his prodigious talents in an attempt to explain an astonishing physical coincidence in cosmology. Comparing the strength of the electrical and gravitational forces acting between the proton and electron in a hydrogen atom, he noticed that the ratio is an immense number, approximately 10^{40} (one followed by 40 zeros). The fact that this value is so large is related to what is now known as the "hierarchy problem." Curiously, Dirac found that the present age of the universe as expressed in atomic units (the time for a light particle to trek across a hydrogen atom) is roughly the same size. In what is known as the Large Numbers Hypothesis (LNH), Dirac suggested that the two numbers are in fact equal.

Sometimes apparent coincidences mask fundamental truths. For example, when physicist Murray Gell-Mann discovered that he could arrange the properties of elementary particles into curious arrays, he speculated that these patterns stemmed from groupings of yet more fundamental objects called quarks. If he had turned out to be wrong, his methods would have been deemed numerological hokum. But he was right, and his insight led to the modern field of quantum chromodynamics—the theory of the strong nuclear interaction that binds protons and neutrons together.

Following a similar hunch, Dirac bet that the coincidence he discovered between various large numbers in the universe stemmed from a fundamental principle of nature. He proposed that the ratio of the strengths of the gravitational and electromagnetic forces was equal in the cosmic beginning but diminished proportionally with each atomic interaction. That is, each time the "clock" of a hydrogen atom ticked, gravity would become slightly weaker. Consequently, by 10^{40} ticks, gravity would be that much scrawnier than still-brawny electromagnetism—the unequal match we witness today. In general form the LNH states that large dimensionless numbers should vary with the epoch of the Universe.*

Is Dirac's result profound or simply prestidigitation? In purest numerical form it is almost certainly not correct, since it does not match up with any known gravitational theory. However, there are compelling ways of altering general relativity to produce a changing gravitational strength that have attracted their share of supporters over the decades.

CHANGING GRAVITY

Spurred by Dirac's curious notion, other physicists have attempted to develop explanations in cosmology and particle physics for why the Newtonian gravitational constant would vary. This parameter, G, an important component of general relativity as well as Newton's law of gravitation, sets the scale of gravity. If G drops in value, the gravitational attraction between any set of masses grows correspondingly weaker.

One of the first such G-variability theories was proposed by German physicist Pascual Jordan in the early 1940s. He recognized that he could modify Einstein's theory by removing explicit references to G and replacing it with a scalar field (representing a spin zero particle). The result is that gravitational strength diminishes over time. Jordan's ideas received little notice at first, likely because of the political situation of the time. During World War II, Jordan was an active member of the Nazi Party in Germany while at the same time generalizing theories developed by an Englishman (Dirac) and a Jewish expatriate (Einstein). This bizarre combination of political and scientific pursuits endeared him to neither side. After the war his theories were finally published, but his record was tainted by his former nationalistic allegiances.

It was not until the early 1960s that Carl Brans and Robert Dicke rediscovered Jordan's notion and granted it more of a hearing. They came upon the idea in a different manner, by developing a novel interpretation of Mach's principle. Brans and Dicke wondered what would happen if G was related to the mass distribution of an expanding universe. Like Jordan, they constructed such an arrangement by replacing G with a variable scalar field. Unlike their predecessor's approach, however, they took extra steps to ensure that energy would be conserved for all times. In Jordan's theory, energy is not conserved. In honor of the developers, the combined proposal is now called the Jordan-Brans-Dicke scalar-tensor theory. The "scalar-tensor" appellation refers to the combination of a scalar field with the geometric tensor of general relativity.

Another model of changing gravity was developed several years later by Hoyle and Narlikar. Similarly based on Mach's principle, it combines the original steady state cosmology with the scalar-tensor theory. It defines the mass of each particle as a field, enabling that value to change from point to point. Thus, electrons could have different masses in different parts of space. This field is a function of the masses of all the other particles in the universe. Because the universe changes, the strength of the gravitational interaction similarly alters from time to time and is also allowed to differ from place to place.

Yet another variable-G cosmology, called scale-covariant theory, was proposed in the 1970s by Vittorio Canuto of the City College of New York. Canuto called for a reframing of Dirac's LNH by means of modifying a number of physical laws, including general relativity and the principle of the conservation of energy. Then, the LNH enters the model as a special condition.

In 1989, Steinhardt and Daile La incorporated aspects of the Jordan-Brans-Dicke theory into extended inflationary cosmology, a variable-G version of the standard inflationary scenario. More recently, a changing gravitational constant has been suggested as a possible solution to the dark-energy conundrum. A diminution in gravity's strength would offer a natural way of explaining the acceleration of the cosmos—a weaker hold allowing for faster expansion.

Researchers have developed numerous tests to distinguish the various contenders for a possible new theory of gravitation and to determine if standard general relativity requires modification. Each of the variable-G models offers specific predictions in the fields of astrophysics and geophysics, consequences that experimenters can readily assess.

ASTROPHYSICAL CLUES

Astrophysics provides us with myriad examples of gravitationally bound systems. Each would be profoundly affected if G happened to vary. Researchers would notice discrepancies on every scale—from individual stars (such as the Sun) to star clusters (such as Messier 67) to galaxies (such as Andromeda) to clusters of galaxies (such as the one in the constellation Hercules) and finally to superclusters (such as the Local Supercluster).

A slow decrease in the gravitational constant would engender a multitude of long-term consequences. Objects in orbit would gradually move farther and farther away from the bodies around which they are revolving. Locally, the average distance between Earth and the Sun would slowly get bigger, cooling our planet over time. Naturally this would be true not just for our own world but for any other planet in space. Hence, if many years in the future we decided to escape our frigid planet, we should expect that any world we settled on would eventually suffer the same fate.

To make matters worse, the Sun and other stars would become less luminous (intrinsically bright) over time. The rate at which a star like the Sun burns hydrogen and other elements depends on its internal temperature, which in turn is governed by the delicate balance of gravity and pressure inside the star. If G decreases, this balance becomes altered and the star becomes less efficient in its fusion. Ergo, it dims like a lightbulb during a power shortage.

With billions of slowly fading stars, galaxies too would gradually lose their luster. New stars would be born, but they'd similarly become less efficient—as the "dimmer switch" of G steadily lowered. Moreover, stars would orbit galactic centers at orbits farther and farther away, making galaxies appear more diffuse. Clusters and superclusters would spread out, in an effect that would add to their existing cosmological recession. In short, the cosmos would become, over the eons, fainter and more dispersed—like the scattered embers of a once-raging fire. (Such a grand cooling process would happen anyway, due to the irreversible law of entropy and the expansion of the universe, but any reduction in G would hasten it.)

We need not wait for eons. There are already a number of astrophysical ways of determining if G has decreased. If the Sun's luminosity has gone down, for instance, meteorites falling on Earth would have been warmer, on average, in the past, affecting the results of their impact and nature of their debris. Thus, meteorite remnants could be investigated with the goal of determining their original temperatures—thereby attempting to find out if the Sun has dimmed over time.

Another astrophysical test of *G*-variability involves the rhythms of pulsars. These dense, rapidly spinning remnants of massive stars emit energy in a periodic fashion. Like lighthouse beacons, they send off pulses as they turn, which astronomers detect as oscillating radio waves (or sometimes X-rays). Gradually, this rate of emission slows, in a process called "spin-down."

In 1968, shortly after the discovery of pulsars, MIT radio astronomers Charles Counselman and Irwin Shapiro demonstrated how their spin-down rate could be used to calculate changes in the gravitational constant. Their paper, published in *Science*, was very much ahead of its time, given that pulsar spin-down had not yet been observed. More than two decades later, at a conference held in Rome, Israeli astrophysicist Itzhak Goldman announced results for the radio pulsar PSR 0655+64. By measuring its spin-down, he placed strong limits on the variability of *G*. He found that *G* changed less than one-billionth of a percent per year.

Although compact objects such as pulsars have extremely high densities, they are able to resist complete collapse into black holes. The bounds of such resistance, set by what is called the Chandrasekhar limiting mass, depend on gravity's strength. Subsequently, if *G* varies, the number of dense stars that end up as black holes would gradually alter. For instance, if gravity grew weaker over time, this rate would plummet. By counting relative numbers of black holes compared to pulsars at various distances (by looking farther out, we see further back in time), astronomers could conceivably detect this effect.

On the grandest scales, a changing *G* would manifest itself in altered patterns of star clusters, individual galaxies, and large-scale galactic distributions. If *G* varied, their gravitational tourniquets would slowly ease up, offering them freer circulation. The result would be a gradual—and potentially discernable—spreading out of these systems.

Taking the Earth's Pulse

In looking for aberrations, sometimes it's wise to start with the familiar. To determine whether or not the gravitational constant is changing, we need look no farther than Earth itself. Aspects of Earth's dynamics, such as its rate of spin, are sensitive measures of gravity's strength. Hence, geophysics provides us with another credible means of testing Dirac's hypothesis and comparing its various incarnations.

By virtue of its rotation, Earth represents a kind of cosmic clock that has been ticking for about 4.5 billion years. During that time it has been gradually slowing down by a rate of about two milliseconds per century. By considering the various processes that could possibly contribute to this lag, we could theoretically deduce information about long-term cosmological effects—such as changes in *G*. In practice, this is a complicated problem because there are numerous mundane effects that contribute to the slowdown. Scientists believe that much of the deceleration stems from ocean tides, caused by the Moon's gravitational attraction. Over time these tugs dissipate energy and gradually decrease Earth's rotational speed.

Newton's laws, applicable to the Earth-Moon system, mandate that angular momentum (the mass of each body times its rotational velocity times its distance from the center of rotation) must be conserved. Hence, as Earth has slightly slowed down, the Moon has compensated by speeding up a bit. This has resulted in the Moon receding, ever so slightly, over the eons. Consequently, by measuring the Moon's orbital motion, we can obtain a precise record of Earth's rotational slowdown, which we can then use to measure any change in the strength of gravity.

There are a number of ways to track the Moon's behavior. The most direct method goes back to the APOLLO project, mentioned earlier, and its predecessors. By beaming a laser pulse to a mirror on the Moon (placed there by astronauts in 1969) and measuring the return time, scientists have developed precise records of the Moon's changing distance from Earth, records that can be compared to what would be expected for constant-G and variable-G models. So far, scientists have found no significant discrepancy from the standard gravitational theory with constant G.

Other techniques for determining lunar motions—and hence ascertaining the rotational history of Earth—involve measuring the occultation (covering up) of stars by the Moon, sifting through records of eclipses (for thousands of years), and examining the growth lines of certain types of fossils that reflect the ancient sequence of days and months. None of these methods are free of ambiguity. However, they have been useful in placing limits on the rate by which G could have varied.

In the early 1970s, Thomas Van Flandern, an expert on the lunar occultation of stars, applied his skills in an attempt to pin down the variability of G. Trained at Yale, he has long taken an iconoclastic approach to the study of the universe—arguing passionately against the standard Big Bang scenario. Using occultation data, he estimated that G has changed by about eight-billionths of a percent each year. This result sparked considerable controversy—to which he is no stranger. More recently, he has argued with NASA officials over his assertions that artificial structures could be seen in photographs of the Martian surface. Despite the controversial nature of his endeavors, Van Flandern has received a number of awards for his work—including one for a prize-winning essay in the Gravity Research Foundation competition.

ATLAS UNBOUND

Apart from long-term changes in terrestrial rotation, one of the curious notions arising from the possibility of a changing gravitational constant, according to at least some of the models, is the one that Earth has expanded over the eons. Today, such a hypothesis has little support in the geological community. If anything, most geologists believe that Earth contracted a bit as it cooled, gained a measure of mass through early meteorite impact, and is currently approximately static. However, in the mid-20th century the expanding-Earth hypothesis, which stated that our planet was once much smaller, was a popular explanation for the appearance of the continents.

Looking at a globe, you can't help but notice that the continents seem to fit together like a jigsaw puzzle. Nevertheless, when Alfred Wegener suggested that the continents were once connected and that they have slowly been drifting apart, he was met largely by incredulity. Until the modern science of plate tectonics (vast segments of Earth's crust sliding over each other) was developed, confirming Wegener's view, researchers had no clear picture of what could cause such immense movements. Sadly, Wegener died in Greenland in 1930 during an abortive mission to measure that island's motion relative to the sea. He did not realize that the ocean floors and continents could move in tandem, floating together on a plastic substratum.

An earlier version of plate tectonics was developed by Australian geologist S. Warren Carey. In the 1950s, when his initial models of continental drift failed to produce the desired results, Carey turned to the idea that an enlargement of Earth would push the continents apart and carve out oceans in the gaps between them. On a globe about half the present size, the continents would have fit neatly together, before gradually separating from each other.

Scientists trying to bolster Carey's theory attempted to identify a plausible energy source. Standard geophysical mechanisms did not seem to work. Due to the long timescale involved, many researchers focused on cosmological causes. However, because Newtonian and Einsteinian theories of gravitation did not provide suitable frameworks, most attention focused on extensions of general relativity that introduced new effects. Theories with a changing gravitational constant, and/or the creation of new matter, seemed to fit the bill quite nicely.

According to these arguments, if gravity's strength has decreased with time, Earth would have become less tightly bound, gradually allowing it to spread outward. The generation of new matter in the universe, hypothesized in theories such as that of Hoyle and Narlikar, would augment this effect. The Hoyle-Narlikar theory predicted an expansion rate of about one-tenth of a millimeter (one-fortieth of an inch) per year. Other models predicted slightly lesser or greater rates, depending on whether or not they involved matter creation. Dicke believed that variable gravitation would produce only a minute expansion, not enough to explain continental drift. Hence, one critical test of theories of changing gravity has been whether geophysical evidence indicated expansion and, if so, at what rate the expansion took place.

In *The Expanding Earth*, published in 1966, Pascual Jordan explored possible connections between cosmology and geophysics. Jordan argued that terrestrial measurements offered an ideal test for particular models of gravitation. He realized this was a tricky business, given the contentious nature of the topic. "We cannot decide immediately," he wrote, "whether the Dirac hypothesis really contradicts experience. . . . We are compelled to inspect the results of various empirical geophysical research and select those whose foundations appear sound, often from controversial statements."

Physicist Engelbert Schucking, who worked with Jordan in Germany on some of his cosmological theories and is currently a professor at New York University, has grown doubtful about the notion that the gravitational constant is changing. "The idea of changing G," asserts Schucking, "is very much refuted by observation—the laser ranging to the Moon. G could not have changed by more than 1 percent from the beginning." He is also dubious about the validity of changing the fine-structure constant and varying the speed of light, two other recently proposed ideas.

With any luck, astrophysical and geophysical data will eventually converge on a consensus for whether or not the gravitational constant varies over time. Earth's solidity, its relationship to the

Moon, and other physical measures seem to place strict bounds on this. Complicating the picture is the possibility that other fundamental parameters could alter in tandem, masking the impact of a single changing value. Thus, one needs to think hard before ruling out any given effect.

VARYING THE SPEED OF LIGHT

Since Einstein's day, physics has embraced the mantra that the speed of light in a vacuum cannot vary. No moving object, according to theory, can exceed this universal limit. This assumption serves as the underpinning for special relativity, which is supported by a wealth of laboratory data. Thus, most physicists take it as sacrosanct that light always moves at the same velocity.

In 1999, theoreticians Andreas Albrecht and João Magueijo, then at Imperial College in London, offered the radical proposition that the speed of light has varied over time. Calling this theory the Varying Speed of Light (VSL) hypothesis, they asserted that it would serve as well as inflation in solving the horizon and flatness problems and could also explain astrophysical data on the cosmological constant.

If signals once traveled from one part of the universe to another faster than they do now, that could explain why space is so uniform. Through a rapid and far-reaching process of thermal equilibrium, temperatures in the early universe would have had ample opportunity to even out. Also, any significant pockets of high- or low-density matter would tend to even out over time through either the release or the accumulation of energy. Although such processes would violate standard conservation laws, they would be permitted if the speed of light could vary. This leveling out would lead automatically to a flat cosmos. Hence, the horizon and flatness issues would vanish, with no inflationary smoothing needed to accomplish these feats.

Furthermore, alterations in the speed of light would affect astronomers' measurements of the velocities of distant galaxies. The supernova results of Schmidt's and Perlmutter's groups, among other measurements, would require reinterpretation. Consequently, the universe might not be accelerating after all (or accelerating at a different rate than previously thought). This would again change our understanding of the cosmological constant.

Magueijo vividly describes how he and Albrecht developed this radical approach. Before they began the collaboration, Albrecht had a "lifelong obsession, the need to find an alternative to inflation." The paper with Steinhardt had been Albrecht's first (it was his doctoral dissertation work), so he felt it was time to examine other possibilities. Together, they explored the changes to the equations of physics needed to realize their idea. These turned out to be quite significant, considering that their theory contradicted not just general relativity but also special relativity. Even long-accepted formulations, such as Maxwell's equations of electromagnetism, required modification. Nevertheless, they ardently pressed on—not knowing where the fruit of their efforts would lead.

Because of the controversial nature of the VSL hypothesis, the researchers had difficulty getting their findings published at first. Journal editors were reluctant to touch material that seemed to challenge the maxims of modern science. It took a year of revisions before their initial article was accepted. Even once their work was in print, many mainstream physicists shied away from it. Soon, this controversy was fueled even further by clashing experimental results pertaining to yet another natural "constant."

ALTERING ALPHA

The gravitational constant and the speed of light are not the only fundamental parameters that researchers have asserted could change with cosmic time. Another popular candidate for variability is the fine-structure constant, known by the Greek letter α (alpha), which is basically the square of the electron's charge, combined with other parameters, including the speed of light. Thus, either a changing electric charge or a changing light speed would cause alterations in alpha as well. Although quantum theory asserts that at ordinary energies alpha remains approximately $1/137$ for all times, physicists have reason to believe it could vary under extreme conditions, such as those in the very early universe.

The fine-structure constant lies at the heart of quantum electrodynamics; it gauges the strength of interactions between charged particles. Because at higher energies, virtual particles arise that shield the charges of real particles, some theoretical models suggest that alpha could have a different value in such regimes. If researchers established that it was not only energy dependent but also time dependent, this would imply a slow change in the properties of the vacuum. Earth's evolution could consequently be affected over long periods. Hence, like changing G, variations in alpha could possibly be detected through geophysics.

In 1999 a team of astrophysicists led by John K. Webb of the University of New South Wales found evidence for evolution of the fine-structure constant in the absorption spectra of very distant quasars, extremely remote, superpowerful sources of energy, believed to serve as the dynamos of young galaxies. Webb and his colleagues found that alpha could have varied as much as 2 percent since the time of the Big Bang.

For those scientists who are of the opinion that at least some of nature's firm footholds are really slippery sands, Webb's results offered the tantalizing prospect of vindication. They seemed to reveal a past landscape significantly different from that of today. The study of "variable constants" kicked into high gear, with an increasing number of researchers eager to explore its exotic terrain. Among these innovative scientists was Cambridge cosmologist John Barrow, who, along with Magueijo and Håvard Sandvik, developed a model of the universe based on changing alpha.

Recent findings, however, have cast doubt on Webb's results. In 2004 a group headed by Nobel Prize–winning physicist Theodor Hänsch reported that its four-year study of atomic emissions uncovered no sign of alpha changing over time. Their experiment was precise enough that it would have revealed a variation in alpha as little as one part per quadrillion per year; nevertheless, alpha didn't blink. Another team, headed by Raghunathan Srianand of the Inter University Centre for Astronomy and Astrophysics in Pune, India, conducted a survey of distant quasars using the Very Large Telescope. Analyzing these quasars' absorption spectra, the team established constraints on alpha variation at least four times stricter than those of Hänsch's group.

Physicists and astronomers continue to examine the stony face of alpha, looking for any signs of a twitch. Probing the widest possible range of objects, from atoms to quasars, they are assessing its sturdiness (or flexibility) with ever-sharper tools. A host of contemporary physical theories await their results.

A MATTER OF SCALE

Each cosmological model rests on the bedrock of particular fundamental principles. Even if certain "constants" actually turned out to vary, other aspects of the cosmos could well transcend time's capriciousness and remain true forever. They need not involve actual physical parameters, such as charge or mass, but might represent simple mathematical rules.

Some modern thinkers, inspired in part by mathematician Benoit Mandelbrot's concept of fractals, have suggested that the universal guiding principle is "self-similarity." Self-similarity, the hallmark of fractal structures, means that a portion of something, sufficiently enlarged, resembles the whole thing. Mandelbrot discovered numerous examples of self-similar geometries in nature—from the delicate patterns of snowflakes to the jagged profiles of coastlines.

Consider, for instance, the shapes of trees. Trees generally have a few main limbs extending from their trunks. From these major branches grow smaller branches; from those, tiny twigs; and so forth.

If you clip off a cluster of branches from the end of a limb and hold it up by its own "trunk," it resembles a miniature tree in its own right. Thus, trees are often self-similar—the parts resemble the whole.

Not that such match-ups would necessarily be exact. Nature yields only approximate fractals. True fractals, with perfect self-similarity, are found only in mathematics. Famous examples are the intricate Koch curve (formed by repeatedly removing the middle thirds of triangles' sides and replacing them with smaller triangles) and the exuberant Mandelbrot set (a lacy design etched out through a special algorithm). Still, the existence of almost scale-free natural structures could reveal critical clues about the hidden architecture of reality's cathedral.

A number of researchers have suggested that the universe itself is a fractal. One of the pioneers of such a hierarchical approach is Robert Oldershaw of Amherst College, who has published numerous papers on the subject. Through comparing the properties of systems on many scales, he "found that there was a considerable potential for physically meaningful analogies among atomic, stellar, and galactic scale systems."

Oldershaw has speculated that nature's hierarchy continues indefinitely—like an unlimited succession of Russian dolls, nested one inside the other. Why assume that galactic superclusters are the highest form of organization in the kingdom of all possibilities? Perhaps, he has suggested, the observable universe comprises but a metagalaxy in a greater realm—a meta-metagalaxy, so to speak. The meta-metagalaxy, in turn, would constitute part of an even larger entity, and so on.

In this spirit, let's construct our own cosmic hierarchy. We divide astronomical objects into seven major classes, covering an enormous range of sizes. The first class includes asteroids, comets, and other types of "minor planetary objects," ranging from several feet to hundreds of miles across. Second come planets, with radii spanning thousands to tens of thousands of miles. Stars come next, with radii ranging from hundreds of thousands to tens of millions of miles. (Note that such gargantuan figures represent living stars; extinguished ones, such as white dwarfs, neutron stars, and black holes are much smaller.) Systems in the fourth class, stellar clusters, have sizes best expressed in terms of light-years. Typically, they are on the order of 10 light-years across. (Recall that a light-year is approximately 6 trillion miles.)

Now let's take a colossal leap and turn to the fifth category—galaxies. These objects are ordinarily hundreds of thousands of light-years across. Clusters of galaxies, the sixth rung on this astronomical ladder, generally contain between 50 and 1,000 galaxies within a region roughly 10 million light-years across. Finally, the seventh level includes superclusters and even larger structures, such as filaments, bubbles, and walls. Superclusters typically have total populations of as many as 10,000 galaxies, housed in a sector about 100 million light-years across. Astronomers used to think they were the largest structures in the universe, until in the 1980s a team led by Margaret Geller and John Huchra of the Harvard-Smithsonian Center for Astrophysics mapped out a three-dimensional slice of space, revealing vast, spongy arrangements of galaxies. In their cosmic map, stringy, bubbly and sheetlike arrays of galaxies—called filaments, bubbles and walls, respectively—bounded relatively empty regions, called voids. The largest structure they found was the "Great Wall," a sheet of galaxies stretching out more than half a billion light-years across.

If nature's operating principle is self-similarity, it behooves us to search for commonalities on all scales. One natural place to look is in the density distribution of various astronomical structures, which indicates how much of their material lies at their centers and how much is peripheral. Clearly, because these systems have mass, the inverse-square law of gravitation constitutes one part of the picture. Additionally, because many of these systems are rotating, their total angular momentum must play a strong role. Moreover, because of the elusiveness of dark matter, we know that the first two factors, applied just to visible material, cannot tell the whole story.

Consider, for instance, the density distribution of a cluster of galaxies. Like a sprawling metropolis, it has a greater concentration of galaxies packed within its central district than it does strewn way out in its suburbs. Hence, at least for visible material, clusters exhibit a density pattern that peaks at its center and drops off with radial distance. Superclusters display similar arrangements among the distribution of their member galaxies. They contain, however, several individual clusters and smaller groups of galaxies. Also, they may not be in equilibrium, meaning their forms are not settled.

Links between density distributions at various scales suggest that relative mass has more meaning than absolute mass in describing the state of the cosmos. After all, absolute mass is but a human invention. When we stand on a scale, we are comparing our bulk to a particular fixed amount. In metric units that standard is one kilogram—originally defined as the mass of a special platinum-iridium cylinder protected in an underground vault in Paris. Surely, galaxies don't stop off in Paris when deciding how to arrange themselves.

Strange as it may seem, by temporarily abolishing the kilogram (and all other mass units) the theory frees up to have no particular scale. Instead, relative mass can be defined as a function of two fundamental quantities—the gravitational constant and the speed of light—as well as of the size of a particular region. This combination of distinct parameters could be an important clue to solving the mystery of why the naturally occurring laws of galaxy distribution comprise but a small subset among all possible arrangements. With these special assumptions in mind, we can construct self-similar cosmological solutions of Einstein's equations that could represent the scale-free organization of material in the universe. By matching them up with the distributions of galaxies in clusters and superclusters, we could explain commonalities between those two scales.

What about smaller scales, from asteroids, meteors, and comets up to planets, stars, stellar clusters, and galaxies themselves? Could there be a simple scale-free rule that would unite these disparate shapes and sizes? At first glance these objects appear as different as can be. Yet each has at least two things in common: gravitation and rotation. Each has mass, and each spins about an axis. Not to say that these masses and spins are at all identical—in fact, they are very different. However, what if there were a simple combination of mass and spin that is itself scale-free? Such a construction would represent a neat way of categorizing the properties of a vast range of astronomical objects.

Out for a Spin

Many amusement parks have rotor rides, best avoided after a hearty lunch. If you haven't eaten recently, let's take a spin on one of these contraptions. After handing over your ticket and waiting in a half-hour queue, you walk into a large metal cylinder. Following the lead of others, you stand with your back against the wall. Soon the contraption begins to rotate—slowly at first, then faster and faster. As the floor beneath you starts to drop, you find your back pinned against the metal surface. Inertia, you realize, as you remember the story of Newton's bucket. If it weren't for the wall, you'd be flying off toward the roller coaster. The surface against your back acts to keep you moving in a circle, exerting what physicists call a "centripetal force," which creates an inward directed "centripetal acceleration," enabling your circular motion. Then, before you can work out the equations, you see the ride operator pull the lever to end the ride. The floor rises, the whirling stops, and you get off. Had fun?

As you disembark, you realize to your dismay that everything still seems to be spinning—from the churning of your stomach to the agitation in your head. The delicate fluid in your inner ear presses uncomfortably upon special nerves, producing the loathsome sensation of dizziness. You feel absolutely queasy, and need to sit down on a nearby bench.

Just as linear motion, in the absence of force, tends to continue indefinitely, rotational motion has its own way of lingering. This kind of persistence is called "conservation of angular momentum." Recall that angular momentum is the product of an object's mass, its rotational velocity, and its distance from the axis of rotation. One outcome of this conservation law is that if a rotating body pulls in its bulk, reducing its radius, it tends to spin faster—the total angular momentum remains constant—so if one quantity, say the distance, decreases, one of the others, say velocity, must increase. Due to conservation of angular momentum, therefore, a pirouetting skater is able to whirl at impressive speeds after drawing her arms closer to her torso. Similarly, pulsars spin considerably faster than the much larger stars from which they evolved.

The conservation of angular momentum is the reason that Earth and the other planets rotate today. The solar system, in its youth, was a whirlpool of gas and dust. As it coalesced into the Sun and the nine planets, they each retained a portion of the original whirlpool's angular momentum—hence, each must also rotate. Similarly, in the amusement park example, if you had a sensitive frictionless gyroscope in your hand while the rotor was moving, it could continue to spin after the ride stopped.

In our search for a scale-free law uniting the cornucopia of astronomical systems, angular momentum provides an important clue. Along with mass, it is a fundamental way of classifying objects in the cosmos. Indeed some objects, such as black holes, are distinguished only by these two parameters (supplemented in certain cases by electric charge).

With the goal of a scale-free principle in mind, we can construct a simple relationship between angular momentum and mass that appears to describe many rotating astronomical systems on a variety of scales. The rule says that angular momentum is proportional to mass squared, at least approximately. Supported by well-known data, the accuracy of this relationship is continually being refined by ongoing observations of galaxies. Peter Brosche, of the University of Bonn, has suggested that a closer approximation involves a power of 1.7 instead of 2. Jagellonian University researchers Wlodzimierz Godlowski and Marek Szydlowski, along with two colleagues from Poland's Pedagogical University, have proposed a slightly more complex rule that includes an additional term. Regardless of the actual equation, all these researchers agree there is a connection between rotation and mass that could well have been set in the inaugural stages of the universe.

Extrapolating this relationship to larger and larger systems leads to the startling prediction that the universe as a whole might be rotating, though data from WMAP and other background radiation surveys place strict limits on this possibility. Nevertheless, as Godlowski and Szydlowski have recently suggested, perhaps a small (hitherto undetected) spinning of the universe may manifest itself as a component of its acceleration. As the rotor example shows, circular motion implies a centripetal force. Maybe rotation offers the universe at least part of the extra push detected in the supernova observations. Although Godlowski and Szydlowski assert that they've found data to support this hypothesis, it remains a highly speculative idea— disputed by many researchers.

The notion that the cosmos is one vast merry-go-round dates back to a 1946 proposal by Gamow, who wondered if "all matter in the visible universe is in a state of general rotation around some centre located far beyond the reach of our telescopes." Three years later the renowned mathematician Kurt Gödel followed up with the first rotating cosmological solution of Einstein's equations. Gödel was quite proud of his result and discussed it with Einstein during their walks together in Princeton.

One curious aspect of Gödel's spinning universe model is that by circumnavigating its rotational axis an astronaut could travel back in time, because time's axis tilts during space-time rotation. With enough tilting, the law of cause and effect would break down, allowing an event from the future to "touch" an occurrence from the past. These causation-violating loops are called closed timelike curves (CTCs).

A good analogy to this idea is a standing circle of dominos, representing time's directional arrows, placed carefully around the edge of a turntable. Suppose that by rotating the turntable slightly one of the dominos would become unstable, topple over, and hit the next, which would strike and tip over the next one, and so on. In short order, all the other dominos would fall, with one pushing over the next in sequence. Typically, the last domino to topple would lie on top of the first. Because the arrows of time would now connect up in a loop, a causal connection would establish itself between the future (the final domino to fall) and the past (the first domino). This would represent a potentially navigable CTC.

So if the universe is rotating and you would like to visit the past, simply hop on a spaceship and take a trip around the axis of spin—the ultimate rotor ride, suitable for those with no sense of vertigo or inconvenient present-day attachments. And unlike journeys into black holes, you wouldn't have to dodge hideous singularities. But if you've already completed your voyage, you knew all this ages ago—or ages from now, as the case may be.

6 Escape Clause: Circumventing the Big Bang Singularity

Naturally we were all there. . . . Where else could we have been?
Nobody knew then that there could be space. Or time either:
what use did we have for time, packed in there like sardines?
I say "packed like sardines," using a literary image: in reality
there wasn't even space to pack us into. Every point of each of us
coincided with every point of each of the others in a single point,
which was where we all were.

Italo Calvino (*Cosmicomics*), translated by William Weaver

TIME ZERO

Pity the hapless soul who comes face to face with a singularity, the nemesis of anyone trying to find a finite, well-defined solution to an equation. Einstein notably despised these mathematical beasts. In the 1930s and 1940s, while working with his assistants on various attempts to unify the various aspects of nature, he counseled that singularities were unnatural, even ungodly. Like a meticulous merchant wrapping up a precious package, he felt that an intelligent creator would not allow open ends. This cautionary note was passed down by Einstein's assistants, such as Nathan Rosen, to their own students. For example, Fred Cooperstock, a student of Rosen's, was used to such admonitions. Distaste for singularities has not faded over time. Today, despite the hard evidence of the background radiation, many researchers still find it hard to accept the idea that the universe was created in a state of infinite density and zero volume. How could it ever get out of such a state?

Almost nobody doubts there was a period in the early universe when it was extremely hot and dense; in fact, WMAP and other measures of the cosmic microwave background provide unmistakable proof. There have been numerous other attempts to account for this radiation, but none of them can explain why its temperature is very nearly the same in all directions. Thus, for want of a better explanation most cosmologists accept that it is the remnant of a primeval fireball.

The debate has to do with the Big Bang singularity itself. Could there be a way of accounting for all observed cosmological results without having the mathematics of the theory go haywire some 13.7 billion years in the past? Could the initial creation of matter be explained by a known physical process, rather than just by fiat?

In the late 1960s, Hawking and Penrose demonstrated that just as black holes must have final singularities, the standard Big Bang must have had an initial singularity. The theorems they proved assumed that the universe contained material of typical density and pressure and that its dynamics could be modeled through ordinary general relativity. Most physicists have accepted these conditions as reasonable and have resigned themselves to a universe of indeterminate origin.

In recent decades, however, researchers have sought ways around this knotty issue. One such proposal was put forth, interestingly enough, by Hawking himself. At a 1981 conference organized by the Vatican, he suggested that space-time has no boundary. By substituting "imaginary time" (mathematically, real time multiplied by the square root of negative one) for real time, Hawking found that he could transform the Big Bang singularity into a smooth surface— akin to Earth's South Pole. Just as Antarctic explorers don't fall off the face of the Earth when they pass the pole, but rather change their direction from south to north, Hawking argued that if someone could travel back in imaginary time, past the Big Bang, they would simply start to move forward in time again. This is all well and good and shows a mathematical way of eschewing the initial glitch. However, to many physicists this explanation doesn't seem *physical* enough, given that nobody can really travel in imaginary time.

Other suggestions for eschewing the initial singularity have been based on quantum randomness. In 1973, Hunter College physicist Edward Tryon published a provocative article in the prestigious journal *Nature*, entitled "Is the Universe a Vacuum Fluctuation?" In quantum field theory, particles continuously pop in and out of the vacuum froth. Tryon pointed out that under extraordinarily rare circumstances a particle could randomly emerge from the foam with the mass of the universe. Though the chances of it coming forth at any given moment would be almost nil, it could literally take all the time in the world to make its debut. Eternity's infinite patience would guarantee its emergence. This quantum fluctuation would serve as the creation event for the entire universe we see today.

Through the theories of Starobinsky, Guth, and others, physicists came to realize that it wasn't necessary for all matter to emerge directly in a single fluctuation. Rather, it could surge forth during an era of negative pressure, spawned indirectly by the fluctuation. Hence, the fluctuation would serve as a catalyst rather than the actual influx of matter itself. For example, a fluctuation that was a scalar field could induce negative pressure—or, the equivalent, a positive cosmological constant—that would inflate a region of space. (Recall how under negative pressure space balloons outward.) Large enough negative pressure would coax matter and energy from sheer nothingness, rapidly filling the universe with material. The fluctuation would therefore seed an inflationary era—or its equivalent.

Guth has pointed out that during the inflationary era mammoth quantities of energy could have emerged from sheer nothingness.

Like the bursting of a colossal dam, more than 10^{88} particles could have rushed into the void—providing the building blocks of everything we see today. He refers to this immense creation of material as "the ultimate free lunch."

This mechanism for the rapid production of matter suggests a two-stage model for the universe: vacant up to a certain time, when a quantum perturbation would trigger an era of negative pressure (inflation), and matter-filled thereafter. The Big Bang would be replaced by a "blip"—a transition between one universal phase and another. Because the early matter-filled cosmos would be extremely hot, there would be sufficient thermal energy to build up the light elements and generate the observed background radiation. Thus, the issues that plagued the steady state cosmology would be avoided.

Another idea, called eternal inflation or the self-reproducing universe, imagines Linde's chaotic inflationary cosmology as an unlimited process, with each universe generating fluctuations that spawn new universes. All that would be needed is an appropriate scalar field emerging in a particular region of space, and a bubble from that sector would blow up into a cosmic offspring. Thus, each generation would constitute the breeding ground for the next.

There is nothing to stop this process from continuing ad infinitum into the future. Could then the Big Bang singularity be avoided by extrapolating this mechanism indefinitely into the *past*— that is, imagining our universe as the prodigal son or daughter of an earlier one, and so forth? Apparently not. In 1993, Borde and Vilenkin proved that even eternal inflationary models would have to start with an initial singularity. Push the moment of reckoning as far back in time as you like, but there still would have to have been a singularity at some point in the past.

UNITING THE FORCES

All of the attempts to avoid the Big Bang singularity discussed so far are four-dimensional in character. Some of the most promising

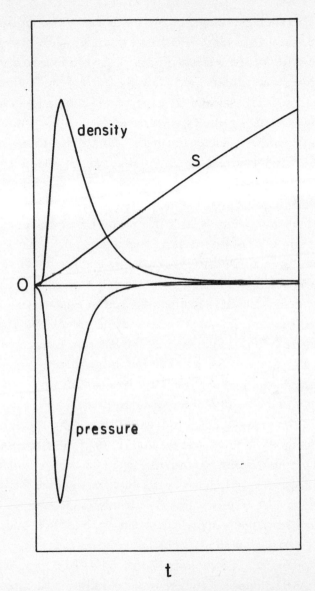

The scale factor as a function of time for a cosmological model that does not start in a Big Bang but evolves instead from a flat (Minkowski) space-time. The density goes through a sharp spike because matter is produced when the pressure becomes negative, in accordance with Einstein's equations. Models like these have been studied by Bonnor and Wesson. They are interesting because they can exist forever, suffer a (quantum) perturbation, experience a period of matter creation, and then evolve into something like what we observe today. (Illustration designed by Paul Wesson.)

modern cosmologies, however, are based on the intriguing notion that the universe has more dimensions than just space and time. The major motivation for these theories is to obtain a unified description of all the natural interactions. Ideally, such a unification would account for the rest masses of all known elementary particles, the interaction strengths of all known forces, and other aspects of the subatomic world. A successful theory would help fulfill Einstein's quest for a single set of equations that would hold the secret to universal dynamics.

The concept of higher-dimensional unification has a long and turbulent history. Shortly after Minkowski first described special relativity in terms of four-dimensional space-time, a young Finnish physicist, Gunnar Nordström, decided to extend these four to five. In his research, published in 1914, Nordström attempted to add a dimension to Maxwell's equations of electromagnetism and thus describe gravitation, the only other fundamental force known at the time, as well. Unfortunately, the gravitational component of his theory turned out to be flawed. When Einstein's successful gravitational theory appeared two years later, Nordström's work crumbled—slipping into the dustbin of forgotten theories.

In 1919, Theodor Kaluza, working as a mathematician in Königsberg, East Prussia (now Kaliningrad, Russia), had more success extending general relativity itself. When he wrote down the five-dimensional equivalent of Einstein's equation, he found that it elegantly divided into two groups of relationships. One set reverted to standard general relativity and the other to Maxwell's formalism. Kaluza's theory thus showed how gravity and electromagnetism as four-dimensional physical forces could be derived from pure, five-dimensional geometry. The theory seemed to wrap the then-known natural interactions up into a single, unified description. Nobel laureate Abdus Salam described Kaluza's reaction as follows: "To his amazement he discovered that he had . . . written down not only Einstein's theory of gravity, but also Maxwell's theory of electromagnetism. . . . It was an incredible and miraculous idea. . . . What

Kaluza did is he sent the paper to Einstein and asked him to get it published."

At the time, Einstein helmed one of the most important journals in Europe, the *Proceedings of the Prussian Academy of Sciences*. Well known throughout the scientific community, he was at the brink of international fame. Very soon thereafter, Eddington's eclipse results would confirm general relativity and boost Einstein to celebrity status.

Kaluza, on the other hand, was at the time virtually unknown. He eked out a living selling tickets to his lectures—a common way that junior professors in Germany functioned until they gained enough status for a permanent position. Fortunately, though Einstein had likely never heard of Kaluza, he was willing to read over the mathematician's results. Although at first glance he found Kaluza's idea extraordinary, he suggested improvements to the paper before it could be published. For example, Einstein hoped Kaluza would provide a sound physical explanation of why the fifth dimension couldn't be observed. Instead, he offered only a mathematical rationale that seemed rather unsatisfactory. Given such a promising but unusual idea, Einstein pressed for a more solid justification.

As Kaluza's son recalled this discourse: "Einstein asked a question or made a suggestion. Then my father did something about it and sent it to Einstein. Einstein asked another question and so on. There are five or six letters of this sort."

For more than two years they corresponded, until Einstein at last decided that the article ought to be published. One of the reasons for the delay, strangely enough, was that Germany was just emerging from the First World War and had a severe paper shortage. Therefore, Einstein felt justified being somewhat picky about the articles he recommended. In 1921 the article finally appeared in print.

Quantum Connections

In 1926, around the same time that modern quantum theory emerged, Oskar Klein, working in Copenhagen, published an intriguing, independently derived variation of Kaluza's five-dimensional theory. In quantum physics, Klein is known for his work on what is now called the Klein-Gordon equation (a special-relativistic extension of quantum mechanics), among other contributions. He was Niels Bohr's assistant during a critical period and thus played a central role in the development of quantum notions. Later he would become a professor at the University of Stockholm, where he continued his theoretical research. He was also a close friend and lifelong correspondent with Gamow, with whom he discussed cosmology and other topics.

Klein translated Kaluza's theory into quantum terms by writing down a five-dimensional variation of the now well-known Schrödinger wave equation. He showed that this five-dimensional equation had solutions corresponding to both gravitational and electromagnetic waves in four-dimensional space-time.

Klein's approach had several advantages over Kaluza's. The first was that the former had a more mathematically rigorous derivation. (Kaluza had taken some shortcuts.) The second was that Klein's model incorporated quantum physics. Lastly, Klein offered a neat hiding place for the fifth dimension. Instead of slipping it under a mathematical carpet, he rolled the carpet up—so tightly, in fact, that no physical observations could be done. Each point in space would comprise a five-dimensional loop, with a diameter less than 10^{-31} inches—far, far smaller than any known elementary particle. Hence, no detector could possibly sense it. The method he used has come to be known as "compactification."

Imagine a garden hose lying on the ground, viewed from a helicopter lifting off nearby. As the chopper ascended, the hose would look less and less like a hollow pipe—with a diameter wide enough to carry water. More and more, it would seem like a straight line, slashing through the field. Ultimately, if the helicopter were high enough, its pilot would have no way of detecting the "extra dimension" of the hose.

Klein's theory, like Kaluza's, turned out to have a major shortcoming. Physicists discovered two additional forces of nature—the strong and weak nuclear interactions. During the late 1930s, Klein attempted to incorporate these forces as well, but his theory received little notice due to the tumult of the Second World War. After the war, the idea slowly made a comeback, picking up momentum during the 1970s and 1980s with the introduction of supergravity and superstring theories. To recognize the contributions of the two original thinkers, all such higher-dimensional methods for unification are now commonly known as Kaluza-Klein theories.

AND THEN THERE WERE 11

One difference between the older theories and the newer ideas is that physicists have come to realize that more dimensions are needed to incorporate all the forces. In 1981, Edward Witten proved that at least 11 dimensions were required to merge gravity with the other interactions. This 11-dimensional theory, called supergravity, seemed to provide a fully supersymmetric description of nature. (Recall that supersymmetry is a hypothesized connection between the two major categories of elementary particles: fermions and bosons.)

The direct motivation for supergravity was a long-standing struggle to find a mathematical description of gravitation consistent with quantum field theory. Attempts to quantize gravity in a similar manner to electromagnetism failed to materialize because of the presence of infinite terms that couldn't be canceled out. Supergravity promised a chance to rectify this situation by proposing that gravity, at ultrahigh energies, forms part of a unified field that can be treated through standard quantum mechanisms. This field "lives" in a world of 10 spatial dimensions plus time. As the temperature lowers—for example, in the expansion of the universe—a phase transition occurs called "spontaneous compactification." Then, seven of the 10 spatial dimensions curl up into compact loops, leaving only three large spatial dimensions—namely, the ones we see today.

When supergravity ran into mathematical difficulties in the early 1980s, theorists flocked to superstring theory as an alternative path to unification. By replacing point particles with minute but finite energy vibrations, superstring theory avoids the issue of infinite terms. The prospect of constructing a completely finite field theory provided welcome relief to bleary-eyed gravitational physicists, exhausted from trying to cancel out infinities.

Like supergravity, superstring theory requires higher dimensions to thrive. The minimum number, as John Schwarz and André Neveu demonstrated, is 10—nine spatial dimensions plus time. Six of them become compact with the lowering of energy. Thus, at ordinary energies the extra dimensions are so minuscule they cannot be observed. To probe such depths it would take an accelerator the size of the Milky Way—unlikely to be built any time soon!

Researchers discovered special configurations, called Calabi-Yau shapes (named after mathematicians Eugenio Calabi of the University of Pennsylvania and Shing-Tung Yau of Harvard), into which these added dimensions could twist up to produce various symmetries. The peculiar topology of each pretzellike figure, especially its number of holes, represents the theoretical properties of a certain symmetry group. Astonishingly, there are tens of thousands of such six-dimensional configurations, offering string theory considerable flexibility.

Theorists hoped the exotic rhythms that superstrings performed as they enacted various modes of vibration would reproduce the properties of familiar subatomic particles—from light neutrinos to massive Z bosons. Like a modern ballet performance, each type of dance would offer a unique representation—capturing the mood (spin, mass, and so on) of a particular particle state. Strings can indeed be very expressive—too expressive, in fact. Not only can they replicate known particles, they can enact the features of myriad entities that have never been seen in nature.

By the late 1980s and early 1990s, researchers were buried under a mountain of excess. There seemed to be too much of everything—too many ways for strings to vibrate, too many types of Calabi-Yau configurations, even too many kinds of string theory itself. As Witten, David Gross, Jeff Harvey, Emil Martinec, and Ryan Rohm of Princeton (the latter four known as the "string quartet") demonstrated, there are five varieties of string theory, each a distinct representation. Classified by their mathematical properties, they have been designated Type I, Type IIA, Type IIB, Heterotic-O, and Heterotic-E.

Heterotic theories deftly combine superstrings with bosonic strings into amalgams that are in certain ways superior to each. (A term borrowed from biology, heterotic means "with qualities better than those of the parents.") In these merged pictures the superstrings travel in one direction and the bosonic strings in the other, like traffic on a two-lane highway, a configuration that offers a natural description of chirality, or handedness, a property possessed by particles such as neutrinos.

The various models include two different categories of strings: open and closed. Open strings are like belts draped across a hanger; their ends dangle loose. Closed strings, in contrast, are like the belt on your waist; they form complete loops. Most known particles can be represented by open strings—with gravitons the notable exception. If all this seems overwhelming, think of the armies of poor young graduate students trying to sort out this mess and make original contributions to the field.

Why five different theories? Could nature be so blatantly redundant? And what of 11-dimensional supergravity? Where did that fit in? By the mid-1990s, unanswered questions called for a new revolution to establish connections between the various models and—researchers hoped—whittle them down to a single, unified theory.

THE SECRETS OF M

About the same time that superstrings were on the roll, assorted theorists, mainly centered at the University of Texas, the University of California at Santa Barbara, and the University of Cambridge, were developing a more general approach called membrane theory. Mavericks included American researchers Joseph Polchinski and Andrew Strominger and British physicists Gary Gibbons, Paul Townsend, and Michael Duff (among many others, too numerous to mention). Extending an idea originally proposed by Dirac, membranes represent particles as flapping multidimensional sheets. These entities can be one-dimensional (like strings), two-dimensional (like conventional surfaces), or indeed any number of dimensions. The only stipulation is that the dimensionality of the object must be less than that of the space in which it resides. Thus, a three-dimensional membrane would be perfectly happy living in a six-dimensional manifold but not on a two-dimensional plane. Townsend dubbed p-dimensional objects "p-branes." Now, many researchers just call them "branes."

For many years, membrane theory was considered string theory's obscure cousin. Few outside the field saw the sense in modeling particles with pulsating sheets if vibrating strings were simpler and would do quite nicely. But then researchers began to find commonalities between strings and branes that encouraged useful connections between various approaches. These links, called dualities, paved the way for what is called the "second superstring revolution."

Many string theorists date this groundbreaking development to a talk by Witten at the University of Southern California in February 1995, where he proclaimed the dawn of M-theory, a smorgasbord of string theory, membrane theory, and supergravity that seemed to include something for everyone. Witten wryly told the audience that the meaning of "M" had not yet been determined. It could represent anything from "magic" to "mystery" to "mother of all theories." Calling it simply "membrane theory" seemed perhaps too restrictive.

Through the wonder of dualities, M-theory brought the five varieties of string theory under a single umbrella. In M-theory, two types of duality come into play. The first, called T-duality, relates certain types of string theory, looped around small circles, with others arranged around large circles. In other words, it ties together two different scales of compactification—the tiny and the potentially observable. The other type, called S-duality, links theories with strong coupling (interaction strength) to those with weak coupling. The latter turns out to be more physically realistic. By applying both kinds of duality, we can mold tightly compactified, less viable models into more acceptable theories with a large extra dimension. Consequently, as Witten and his colleagues discovered, linking the string theories implies an 11-dimensional framework that includes the three dimensions of space, the dimension of time, the six curled-up dimensions of the Calabi-Yau space, and the extra dimension produced through duality mechanisms.

Let's set aside the six-dimensional Calabi-Yau geometry (twisted up beyond detection) for the moment, and focus on the remaining five: space, time, and the extra dimension. That way, we'll concentrate on what we can measure. According to M-theory, this fifth dimension does not have to be tiny; it can be indefinitely large in size.

How can this be? Recall the tale of an ant crawling on the surface of a basketball. Because it is constrained to be on the ball's surface, it does not care about the distance to its center. Similarly, a fifth dimension could exist for which we do not have access. Therefore, it could be as large as the other dimensions but wholly undetectable, except perhaps for its impact on gravitation.

The idea that the fifth dimension could be comparable in scale to the others dates back decades. As we'll see in Chapter 7, it underlies induced-matter theory, first proposed in the 1980s. However, M-theory helped focus the attention of the physics community on this concept. M-theory pictures this extra dimension as forming a gulf between two three-dimensional spaces—with only gravity able to navigate the gap. The two "shores" are known as Dirichlet p-branes, or D-branes for short, and the sea in between is called the bulk. For mathematical reasons, open strings—representing almost all known particles—must forever stick like barnacles to the

D-branes, while gravitons, as closed strings, are free to swim through the bulk. Thus, composed of closed strings, we remain perpetually landlocked—with only our gravitational emissions able to plunge beyond our brane.

Because of gravity's special ability to escape, its immersion in the bulk would effectively dilute it. The larger the bulk, the less contact with our brane it would have and the weaker it would appear. Gravity becomes the puny partner of the other forces.

String theorists soon realized that the relative weakness of gravity was one of M-theory's strengths. From the time of Dirac, researchers have sought a solution to the question of why the other three subatomic forces are so much more powerful than gravitation. With the hope that M-theory would resolve this riddle, several groups set out to construct "brane-world models": interplays of bulk and brane designed to replicate precisely gravity's distinctive behavior.

Brane Benders

In 1998 a team of Stanford physicists published one of the first and simplest brane-world scenarios. Known as the "ADD model," for the initials of its designers Nima Arkani-Hamed, Gia Dvali, and Savas Dimopoulos, it offered a bold attempt to resolve the hierarchy puzzle and other issues. Remarkably—for the abstruse world of string theory—it stuck its neck out with clear, testable predictions.

According to the ADD scenario, everything we see in space— visible galaxies, quasars, and the like—resides on a D-brane. Separated from our brane by roughly a millimeter (1/25 of an inch) is a second shadow realm. In between, like the filling in a sandwich, is a thin layer of bulk, accessible only by gravitons.

The ADD team chose that particular thickness of the bulk to model the actual weakness of gravity compared to the other forces. Too much filling would create an indigestibly large discrepancy; too little would not produce enough of a bite to provide a distinction. Even for the best matching case, the researchers realized that their scenario would slightly modify the long-established law of gravity. Such a minute difference could be picked up, however, only at distances comparable to or smaller than the bulk thickness—that is, at one millimeter or less. Since at the time they made the prediction, gravitational measuring instruments had not yet probed such tiny ranges, the researchers felt free to make such an assertion. Within a few years, however, Eric Adelberger of the Eöt-Wash group used ultrasensitive torsion balance experiments to rule out such deviations down to scales much smaller than a millimeter. These investigations placed sharp restrictions on the model, and theorists await the results of further testing.

One variation of the large extra dimension scenario, called the "manyfold universe model," offers an intriguing possible explanation for dark matter. Developed by the ADD group, along with Stanford researcher Nemanja Kaloper, it posits that the visible universe resides on a single brane, folded up like an accordion. In between each crease, slivers of bulk preclude light from passing through. Photons, after all, are open strings and must cling forever to the brane. Gravitons, on the other hand, can freely jump from one fold to the next. Thus, in this model, gravitation reaches beyond where luminous radiation can penetrate. For example, imagine that a star or galaxy is located on the next fold over from ours. Because its light rays would need to travel a long distance along the brane to reach us, it would appear extremely remote—or, perhaps, not even visible at all. Yet its gravitons could jump across the thin layer of bulk and influence a part of space much closer to us—the Milky Way's halo, for example. They could slightly warp that region, leading to a gravitational lensing effect. The result would be the phenomena we associate with dark matter. Hence, according to the theorists, what we call dark matter could well be visible material situated on another fold of our brane.

As we've found with many theories, in battling one cosmological mystery, it's tempting to try to vanquish them all. While attempting to exorcise the dark-matter demon, the ADD group tried to slay the horizon problem as well—without applying the broad sword of inflation. The idea is that the universe, once twisted up into a tight space, has since opened up (along at least one dimension)—like origami restored to its original sheets—thereby separating regions that were originally in close contact, leading to large-scale uniformity. As the group wrote: "The folded universe picture permits apparently superluminal communication between different segments of the brane through the bulk. This could give a non-inflationary solution to the horizon problem, if the brane was originally crumpled in a small higher-dimensional box and later unfolded."

Competing with the ADD scenario and its "manyfold" variation are other brane-world types with fundamentally different properties; chief among them are the models (mentioned earlier) of Lisa Randall and Raman Sundrum. While one of these types has the standard two branes delineated by M-theory, another possesses but a single brane with a warped infinite extra dimension. The warping refers to the bulk having a nonflat five-dimensional geometry—namely, anti-de Sitter space, which serves as a trough, focusing the gravitons in a region close to our brane. Hence, gravity is weak but not too weak.

The Randall-Sundrum model can be pictured as an endless desert with a giant rock in the middle—akin to Uluru (Ayers rock) standing tall in the Australian wilderness. Uluru's location represents our brane; the desert stands for the bulk around it; and the rock itself, all matter and energy besides gravitons. Naturally, the rock remains fixed to the site, like conventional matter on the brane.

Now suppose that a desert spirit suddenly transforms the rock into a giant block of ice, like a glacier. This picture represents gravitons. If the desert is completely flat, the ice would quickly melt, then spread out over an extremely large area. This thin coating of water would rapidly evaporate. By analogy, gravitons exuding into a flat, infinite bulk would have absolutely no strength. But if the desert around Uhuru were slanted toward the center, it would collect the water into a substantial pond. Similarly, a warped brane would localize gravitation—producing a weakened but still significant force like that which we actually observe.

COSMIC COLLISIONS

The unusual designs of the ADD, manyfold and Randall-Sundrum models seem to call out for a study of their dynamics. Following the well-trodden path of Einstein, Eddington, Hoyle, de Sitter, Gamow, and others, it would seem natural to explore the cosmological implications of these theories, pushing them forward into the future and backward into the past. Many string theorists have certainly considered this exercise. As Sundrum, for instance, has remarked:

> When I find a model of physics or a theory of physics that I find particularly attractive for other reasons, then I think it's intriguing to go and study its cosmology. For example, the standard model of particle physics is certainly a very well motivated theory, backed by experiments and certainly it's a good thing to study the cosmology associated with the standard model very seriously.

That said, Sundrum quickly adds:

> I have not yet found theories with brane cosmologies so attractive that I want to directly study the cosmology. I don't find it personally a robust enough activity that points towards the answers that I'm looking for. But I'm very happy that other people do engage in this, because it may turn out something more robust than I had anticipated.

One of the most active groups studying brane cosmologies includes researchers from Princeton and the University of Cambridge. With Gibbons, Townsend, and Neil Turok on staff, Cambridge has become one of the foremost international centers for membrane theory research. Appropriately, these researchers concoct their odd-shaped multidimensional creations within a building of similarly unusual geometry, the Centre for Mathematical Sciences. With curious, grass-covered domes and striking triangular towers, it resembles the futuristic set of a science fiction epic or of the children's television program *Teletubbies*. If a sign was posted that read "Portals to other branes inside," we would not be all too surprised. Ironically, the structure lies on the outskirts of one of England's most traditional university towns.

In a fruitful transatlantic collaboration, Steinhardt, Turok, and several other theorists—including Justin Khoury, then at Princeton, and Burt Ovrut of the University of Pennsylvania—proposed two related cosmological models involving colliding branes. These two cosmologies differed mainly in their time frames. While the first, called the Ekpyrotic (renewed by fire) universe, was a one-time smash-up, the second, called the Cyclic universe, repeated its sequence indefinitely. In either case, enormous quantities of energy would be produced in the crash, enough to represent the output of either the Big Bang (in the original standard model) or the end of an inflationary phase.

In codeveloping these theories, Steinhardt decided to take a break from his quest for a seamless inflationary scenario. Although he still valued inflation, he thought it important to explore alternative descriptions of the universe, including ones involving strings and branes. As he explained: "I've been waiting for string theory to get to a point where there was some kind of loosely speaking phenomenological model you could begin to think about more seriously. Then I asked the question, could you do something interesting with it that would be a different kind of cosmology?"

In the Ekpyrotic model, the universe begins as a standard M-theory solution, with several three-dimensional branes immersed in a higher-dimensional bulk. One of them is the "visible brane," representing our own spatial enclave. Another is the "hidden brane" signifying the shadow world suggested by M-theory. At first our visible brane bears scarce resemblance to the matter and energy-filled space with which we are familiar. Rather, it is a cold and empty place—as welcoming as an icefield in Antarctica.

All this would remain frozen, if it weren't for a third mobile brane—a cosmic daredevil. Like a reckless racer, it moves through the fifth dimension and crashes head on into our brane. The wreckage produces a blazing inferno of energy and matter, distributed throughout our space. Thus, in contrast to the standard Big Bang model (which starts off infinitely hot), the observable universe begins its life at a finite temperature.

The fact that the moving brane hits ours all at once suggests a noninflationary resolution of the horizon problem. A common cause—the collision—explains large-scale uniformity. Additionally, quantum fluctuations scattered throughout the moving brane generate the seeds of structure as that brane touches ours. Thus, according to the Ekpyrotic researchers, neither a Big Bang nor an inflationary era would be needed to explain what we see around us.

CYCLES OF FIRE AND ICE

Many ancient cultures consider cosmic time as renewable and repetitive as the rising and setting of the Sun. It's natural to wonder—whether philosophically or scientifically—if the Big Bang was not the beginning, then what grand eras preceded our own? Could there have been exotic worlds and advanced civilizations that vanished in the fires of a cosmic transition?

Tolman's oscillatory model, proposed early on in the history of general relativity, bore some resemblance to traditional cyclic schemes. However, because it accumulated entropy (disorder) from era to era, it was not truly renewable. Thus, it could not produce an eternal succession of viable cosmologies. The Cyclic universe, proposed by Steinhardt and Turok, attempts to address these issues while explaining the origin of dark energy and resolving other cosmological questions. Like the Ekpyrotic model, it involves colliding branes—in this case bouncing off each other.

Let's run through the cycle that Steinhardt and Turok proposed. First, the branes collide, essentially wiping out all traces of what existed in the universe before. The resulting burst of power replicates what we construe as the Big Bang but with no singularity. Quantum fluctuations in the impacting brane seed the formation of structure in observable space—the galaxies we see today. As the colliding branes move apart, the vacuum energy of the universe changes. The resulting dark energy produces universal acceleration—driving the galaxies farther and farther apart. That's the phase we're in today. Then, as the universe evolves, it will exhaust its usable energy. With stars dying out—turning into white dwarfs, neutron stars, and black holes—entropy will build up more and more. However, as galactic recession speeds up, this measure of disorder will become more and more dilute. In due course, for any given region, it will effectively revert to zero.

Meanwhile, the branes will eventually reverse course and come together again. The visible universe will cease its acceleration and begin to slow down, heralding the calm before the storm. Then, there will be unmistakable harbingers of doom. As Steinhardt describes these signs:

> Once the universe turns to the decelerating phase you still have roughly another 10 billion years to go. But finally, in the last few seconds you'd see some significant changes in the fundamental constants and that would be the hint that something is about to happen. You would notice that something really strange is happening in the universe. Some tremendous form of energy is building up all of a sudden. It would reach a crescendo, and then, bam, the universe would fill with matter and radiation. That would be the collision. You and I would be vaporized unless we were otherwise protected.

Black holes would survive, but most things would be vaporized. So then that universe is full of matter and radiation again.

All earthly civilization would cease. There would be nowhere to escape. However, as Steinhardt relates, you could try to contact any intelligent souls in future eras, providing them with some inkling that our culture once existed:

> You might be able to send messages, although [respondents] would be pretty rare. A black hole survives, so you could make an arrangement of black holes that spell out 'Hello.' The problem is that during this period of accelerated expansion, the universe has expanded exponentially. So the only people able to read that message are people right near where you were. That's a very small fraction of the total population of the universe. We can only see 14 billion light-years today, so the chances of communicating with civilizations spread out [to the extent of] maybe one every hundreds of trillions is negligible. So you can only send messages to your local neighborhood at best.

Unlike the Tolman model, this sequence of events could repeat itself forever, because the accelerating phase dilutes the entropy for a given region, before it fills with new matter and energy. It serves as a conveyer belt to remove the scraps from the table, before new helpings are served. What an efficient cosmic cafeteria.

Despite their clever attempt to use M-theory to resolve cosmological issues, Steinhardt and Turok's model has been met with skepticism by string theorists as well as cosmologists. Many string theorists believe that the model isn't ready to be used in a dynamic description of the evolution of the cosmos. Many cosmologists, on the other hand, ardently believe in the inflationary model, the anthropic principle, or other longstanding resolutions of the horizon and flatness problems.

Many mainstream physicists and astronomers simply aren't used to dealing with five or higher dimensions. Yet we should point out that the fourth dimension was controversial for a number of years after Minkowski first identified it as time. Even Einstein needed to be convinced. Eventually, he and others came to accept the four-dimensional nature of space-time.

Just as the fourth dimension provided a natural way of describing time, introducing the fifth dimension can do the same for mass. It provides a natural way of resolving one of the fundamental mysteries: How did all the mass in the universe arise?

7   What Is Real?

Why, sometimes I've believed as many as six impossible things before breakfast.

Lewis Carroll (*Through the Looking-Glass*)

WEIGHTY GEOMETRY

A baker would be astonished if every cookie he baked had exactly the same size, within thousandths of a millimeter. A sculptor would be amazed if every clay urn she hand molded had precisely the same shape. Yet nature's artisan seems to have crafted untold quantities of protons (and other elementary particles) with identical rest masses. They are infinitely more "cookie cutter" than anything in a cookie manufacturer's wildest dreams.

Why is the mass of a proton on Pluto the same as that of a proton in Pittsburgh? How do they "know" how to coordinate their attire, like soldiers in a vast army? And where does their mass originate anyway? Mass is a fundamental feature of objects in the cosmos. Any comprehensive model of the universe ought to explain how it arises and why it is doled out in identical amounts.

Let us start by examining theories of gravitation. Given the essential relationship between gravity and mass, one might presume that an explanation of one would also account for the other. Not necessarily so. In general relativity the force of gravity results from the curvature of space-time by matter. But it does not explain where the mass in the universe comes from. Einstein's theory merely assumes that the mass came into existence when the universe came into being and has since remained unaltered.

Einstein himself was dissatisfied by the dichotomy between the vibrant, flowing geometries on the left-hand side of his gravitational equations and the stultified stuff on the right. The mass terms seemed positioned on the right-hand side as necessary but awkward ballast. Like the counterweights on an elevator, they helped lift the geometric side to ethereal heights. It would be nice, Einstein felt, if everything were lofty and dynamic and there were no need for extra bulk. As he wrote in an essay, "Physics and Reality":

> [General relativity] is sufficient—as far as we know—for the representation of the observed facts of celestial mechanics. But it is similar to a building, one wing of which is made of fine marble (left part of the equation), but the other wing of which is built of low-grade wood (right side of equation). The phenomenological representation of matter is, in fact, only a crude substitute for a representation which would do justice to all known properties of matter.

Thus, one of Einstein's principal goals in the latter half of his life was to perform the alchemy of turning wood into marble. His motivation for this effort stemmed largely from his strong belief in Machian ideals. Mass, Einstein felt, should draw its nature from the relationships between all objects in the universe. Pure geometry would be the proper mechanism for conveying such information.

Although Einstein first explored such notions in the 1920s, they were hardly new. In the early 1870s the British mathematician William Clifford caused quite a stir with his proposal that matter represents lumps in the carpet of space. His article, "On the Space Theory of Matter," postulated that empty space is completely smooth and that deviations from such smoothness manifest themselves as physical materials. Since space is three-dimensional, these bumps would need to poke out along an additional dimension. Decades before Minkowski, Clifford based his theory on space, not space-time. Hence, the extra dimension of his theory was the fourth. Today, we'd call it the fifth, including time.

Clifford's radical conception of mass was debated for years in the pages of *Nature* and other journals. Readers pondered ways to fathom his multidimensional vision. Sadly, he had little time to develop it further. Pulmonary illness took his life in 1879, when he was only 33.

Although neither Clifford, nor Einstein, succeeded in geometrizing matter, many others have since tried their luck. Like Brigadoon, the shining city of marble has periodically reawakened—enticing eager adventurers to explore its beauty. In the 1950s and 1960s, for example, John Wheeler attempted to describe all particles as geometric twists, called "geons." Think of geons as whirlpools arising, moving, and interacting in the ocean of space-time connections. Recently, physicists Dieter Brill, James Hartle, Fred Cooperstock, and others have revived Wheeler's notion—attempting to stir up various material configurations from the froth of gravitational waves. Finding a mathematical explanation for the solid forms around us is a vision too vital to ignore.

INDUCED MATTER

New five-dimensional cosmologies suggest an intriguing way of achieving Einstein's dream of describing matter through geometry. Let's consider a variation of Kaluza-Klein theory in which the fifth dimension is not compact but rather of observable proportions. We've seen such an assumption put to good use in brane-world models, leading, for example, to the Ekpyrotic and Cyclic cosmologies. While such models account for differences between the properties of various interactions, they do not furnish a geometric explanation of mass's origin. Rather, they introduce matter (in terms of strings) as an extra element—not intrinsic to the universe itself. There is, though, an alternative way of approaching higher dimensions, which naturally incorporates matter into the structure of the cosmos itself. In "induced matter theory," the fifth dimension turns out to be mass. Space, time, and matter comprise an inseparable whole.

Induced matter theory and membrane theory are similar mathematically but clearly different conceptually. One of the main distinctions between the two theories is how they handle energy. This might be the garden-variety type of energy present in the rest masses and motions of particles or exotic forms associated with the "vacuum" (which we know is not really empty). In induced matter theory, the starting equations look like those for five-dimensional empty space but break down naturally into four-dimensional relationships that correctly describe matter and its fields. Hence, the fifth dimension is all around us: It is the energy of the world, whether in the rest masses of particles, the kinetic energy of their velocities, the potential energy of their interactions, or extra contributions involving what has traditionally been called the vacuum.

According to membrane theory, in contrast, the extra parts of the manifold are not apparent to the eye. Rather, like the Wizard of Oz, they operate behind the screen. Furtively, these unseen regions control the gravitational interactions of particles and therefore ultimately the matter of everyday existence. Thus, unlike induced matter theory, what you see is not what you get.

Let's examine the procedure in induced matter theory by which geometry can give birth to mass. In the manner of Kaluza and Klein, we first extend conventional general relativity by adding an extra dimension. We make sure to use the vacuum version of Einstein's theory, with no explicit source terms (matter and energy put in by hand). This ensures that no slivers of wood are tracked into our elegant marble foyer.

We also insist that the fifth dimension remain noncompact. Unlike Kaluza, we don't mathematically dismiss it and, unlike Klein, we don't curl it up. This creates a bevy of extra terms on the left-hand (geometric) side of the generalized form of Einstein's equation. These turn out to be a blessing, not a curse. By moving these to the right-hand (matter-energy) side of the generalized Einstein equation, we can generate matter and energy terms. We can identify these extra quantities, for example, as the density and pressure of actual materials.

Now, let's stand back to see what we have created. On the left-hand side of the equation, we witness the geometric part of ordinary four-dimensional general relativity. On the right-hand side, we find expressions for matter and energy. Where has the fifth dimension gone? Instead of being wrapped up in a cocoon, it is flitting around as the wondrous things we see in space. What a stunning metamorphosis indeed.

If higher dimensions are merely inventions, they are uncannily clever ones. They offer ways of encapsulating complex aspects of nature into simple expressions. Now we see that they can replicate matter and energy. If something looks like reality and acts like reality, perhaps it is reality. A hidden diamond unearthed and polished, the fifth dimension could well be authentic.

MAPPING OUT REALITY

If matter in the cosmos arose through a fifth dimension, how could we best envision the mathematics behind this novel concept? Our brains are ill equipped to accommodate realms beyond the perceptual delimiters of length, width, breadth, and time. How could we best become familiar with the roads and byways of a five- (or higher) dimensional topography?

The answer is that we need a map. Specifically, we need a way of rendering a portrait of five-dimensional interactions in four-dimensional space-time. The latter, in turn, naturally breaks up into a three-dimensional ordinary space, as well as time. We know that on large scales the three-dimensional space is flat because of WMAP and other measures of the cosmic microwave background. These considerations of uniformity and spatial flatness turn out to be very lucky in our quest to map out the universe. They make it easier for us to embed (represent in a higher-dimensional space) the universe in a five-dimensional manifold that is also flat. This picture helps us visualize directly the universe's geometry.

Mathematicians distill lower-dimensional images of higher dimensions through either slicing or projection. Slicing involves carving out lower-dimensional segments like a butcher thinly divides up ham. Thus, cubes can be sliced into squares, hypercubes into cubes, and so forth. In any given slicing, connections are broken, and certain types of information can be lost. Projection, on the other hand, attempts to grasp the whole picture at once. This is done by a process akin to the creation of shadows. For example, by shining light on a hollow cube, we can explore its image on a flat screen. (We can readily see why flat screens would be easier to handle than curved ones.) Similarly, we could imagine illuminating a hypercube to see how it would appear. Such a projection provides a "map" of the higher-dimensional surface.

Constructing such a map is not as simple as it sounds. To see that the problem is nontrivial, consider the map maker's task of rendering the curved surface of Earth onto a flat page. The Mercator projection, used in many school atlases, is very useful for this purpose. However, it distorts the areas of land masses, making regions near the poles appear larger than those near the equator. (It is said that the 19th century British liked this mapping because it exaggerated the size of Canada, Australia, and other parts of their empire.) Equal-area projections, used by many geographers, address this problem but look rather odd. There are indeed an infinite number of ways of making a map, either for Earth or the universe. The value of a map depends on how it will be used.

To visualize a higher-dimensional cosmos, we have a clear plan. First, we express flat three-dimensional space as part of a curved four-dimensional space-time. Then we embed the latter in a flat five-dimensional manifold. The mathematical justification for this arises from a theorem proved by the early 20th century Irish geometrist John Edward Campbell. Campbell's theorem states that a certain simple multidimensional manifold can properly represent any surface within it of one fewer dimension. Consequently, the five-dimensional equivalent of a plane can well house any four-dimensional occupants—no matter what their shape or size.

Because we are trying to represent the real universe and not some hypothetical construct, we have to ensure that we do not contradict the physics involved. The physics is encapsulated in field equations that relate to particular equations of state.

Recall that an equation of state defines the precise connection between the pressure and density of a material. Radiation, dust (loose material), and tightly packed matter each have different relationships, reflecting varying types of movement in response to forces. When the universe was very young, its particles were extremely energetic, and its equation of state was close to that of photons, where the pressure is one-third of the energy density. In the present epoch the energy in the microwave background is many orders less than that in galaxies and dark matter. This means that currently the equation of state is analogous to that of dust; that is, pressure effectively equals zero.

According to induced-matter theory, the properties of three-dimensional matter, evolving in time, arise as a "shadow" of five-dimensional geometry. Hence, the universe's equation of state during various epochs stems from specific relationships between sets of geometric terms in the five-dimensional extension of general relativity. These extra terms result from an uncurled fifth coordinate in Kaluza-Klein theory.

In 1988, Jaime Ponce de Leon, a young theorist from Puerto Rico, solved the five-dimensional field equations for a noncompact fifth dimension. His results were most remarkable. Finding the pressure and density relationship between various classes of solutions, he discovered that they precisely matched the known equations of state for specific types of matter and radiation during different universal epochs (inflation, radiation era, and so on). In other words, shadows of the fifth dimension had profiles similar to those of the familiar faces of cosmology. This impressive correspondence boded well for the theory.

With reasonable solutions in hand, we can now map the terrain. Curiously, the relationship between these four-dimensional profiles (technically known as hypersurfaces) and the five-dimensional world that surrounds them resembles that of blobs in a 1960s lava lamp. Imagine a blob of oily material floating in water. The physical properties of the blob's surface determine its changes in shape over time. Similarly, the physical characteristics of four-dimensional space-time determine its dynamic relationship with the five-dimensional manifold in which it is embedded.

One striking feature culled from this exercise in higher-dimensional cartography concerns the shape of the Big Bang. While in four dimensions the Big Bang has an unavoidable singularity, in five dimensions the singularity vanishes. Rather, the five-dimensional topography is as smooth as crystal. No blemish marks the initial burst of the universe—permitting a fuller and more satisfying description.

Another fascinating result of these studies could bear on the dark-matter question. Particular solutions comprise durable mathematical structures known as solitons. If five-dimensional solitons exist, they could well provide an important piece of the puzzle of why so much of the material in space cannot be directly observed.

SOLITONS FROM THE DEEP

The curiously persistent forms called solitons were discovered by the Scottish engineer John Scott Russell during a survey of boats floating along a canal. While observing a barge's motion, he noticed "a rounded, smooth and well-defined heap of water, which continued its course along the channel apparently without change of form or diminution of speed. I followed it on horseback, and overtook it still rolling on at a rate of some eight or nine miles an hour, preserving its original figure some thirty feet long and a foot to a foot and a half in height. Its height gradually diminished, and after a chase of one or two miles I lost it in the windings of the channel. Such, in the month of August 1834, was my first chance interview with that singular and beautiful phenomenon which I have called the Wave of Translation."

What Russell witnessed was a particular phenomenon that can appear in a shallow body of water. The dynamics of such a system, governed by what is now called the Korteweg-deVries equation, permits "solitary waves" that do not diminish in amplitude or spread out in space as they move along. Such solitons spontaneously appear if hydrodynamic conditions are just right.

Since Russell's time, solitons have assumed an important role in topology and other branches of mathematics. A number of noteworthy equations have distinct soliton solutions that do not dissipate over time. Rather, they maintain their shape indefinitely as they propagate. In physics, solitons have offered hope for representing particles as "kinks" in the fabric of field theories—in a manner akin to Wheeler's geon model. By taking the well-known Klein-Gordon equation and replacing a term with the sine function, the result is the "Sine-Gordon equation," which produces soliton waves in quantum physics.

Given that Kaluza-Klein theories harbor many modes of behavior, it is not surprising that among these are five-dimensional solitons that represent nondissipative solutions in the induced-matter scenario. Discovered in the early 1990s, they constitute higher-dimensional generalizations of the Schwarzschild model.

As noted, the Schwarzschild solution, published in 1916, describes the properties of nonspinning black holes of neutral charge.

In particular, it specifies at what particular radii their event horizons lie—as determined by their masses. As far as researchers know, all black holes have event horizons that serve the vital function of clothing points of indeterminate properties—in other words, singularities. Otherwise, such singularities are said to be "naked."

Compared to the Schwarzschild solution, five-dimensional solitons are quite different beasts altogether. Each represents a concentrated, spherically symmetric form of induced matter. Unlike ordinary, pointlike black holes, typically these clouds of material are extended objects with densities that sharply decrease with radius. Moreover, their pressures are generally anisotropic—different in each direction. They possess equations of state characteristic of ultrahigh-speed particles. Because we have yet to predict the properties of their light spectra, we do not know what range of temperatures they could have. Like unmarked faucets of energy, they could run hot or cold.

One of the most important distinctions between these "fireballs from the fifth dimension" and ordinary black holes is that the former do not possess event horizons. No space-time garment covers up their singularities. If they exist, they are astronomical streakers, displaying their entire selves for any telescope powerful enough to see them. However, it's quite possible, if the five-dimensional solitons are cold enough, that they'll emit little-to-no discernable radiation. In that case, only indirect means—such as their interactions with visible stars, their influence on the development of galaxies, or their gravitational effects on passing light rays—would potentially distinguish them from the void. This third means, gravitational lensing, would likely provide the best opportunity for finding them. Astronomers could probe for patterns in the emissions of quasars, distant galaxies, and other objects, as distorted by the unseen presence of intermediate bodies. Then they could match these results to soliton profiles.

If Kaluza-Klein solitons turn out to be plentiful enough, they would be prime suspects for the hidden material that fills the universe. Moreover, five-dimensional theories also allow for other kinds of matter and predict that space is permeated by a kind of vacuum field. The latter arises directly from the scalar field connected with the fifth dimension and acts like a variable cosmological "constant." Thus, the five-dimensional theory agrees with the COBE and WMAP observations in describing the universe as consisting of ordinary (visible) matter, dark matter, and dark energy.

BURGEONING MASS

One of the curious features of induced-matter theory is that mass has the same units as length. At first glance this would seem strange. If you took your child to the doctor's office for repeated check-ups, you would be perplexed if the tape measure and scale always read the same. Remember, however, that the length corresponding to mass extends along the fifth dimension and is independent of three-dimensional space. Thus, a physician could delicately report to the mother of an overweight child: "My, your son has grown—even in the fifth dimension."

Physicists perform a similar conversion in standard relativity when they convert time into distance using the speed of light. In the metric system the speed of light has units of meters per second. Hence, multiplying time (in seconds) by the speed of light "magically" converts it into distance units (meters). This procedure provides time with the proper membership card to join the club of spatial dimensions as its fourth member. The conversion factor that transforms mass into distance and allows it to become the fifth member of the dimensionality club is the ratio of the gravitational constant to the speed of light squared. That is, if you combine the units of these parameters with those of mass (kilograms), you end up with distance units.

Many theoreticians set various constants equal to one to simplify their calculations and make the math more readable. By setting the gravitational constant and speed of light each equal to one, mass and length now have precisely the same units. They simply constitute perpendicular ways of measuring the world—the former along the fifth dimension and the latter along any direction of the spatial threesome. This lends itself to an intriguing reinterpretation of the notion of a changing gravitational constant. Current tests, such as laser measurements of the Earth-Moon distance and investigations of the Sun's luminosity, place strict limits on alterations in the strength of gravity as the universe ages. However, this force depends on objects' masses as well as the gravitational constant, G. What if the masses of elementary particles are themselves changing over time?

Brans and Dicke pointed out in their seminal paper that, if you construct a representation of mass in length units (ordinary mass times the gravitational constant divided by the speed of light squared), it is natural to imagine that this quantity would grow with the size of the Hubble radius (the boundary of the visible universe). Just as distances between galaxies increase with time, according to this hypothesis, masses would as well. This is completely equivalent to the theory of varying G. Instead of mass being constant and G variable, mass would alter and G would remain constant. As Brans and Dicke emphasized, "There is no fundamental difference between the alternatives of constant mass or constant G."

According to the induced-matter hypothesis, mass derives from dynamic solutions of the five-dimensional extension of general relativity. Therefore, it is not surprising that solutions exist in which particle masses vary slowly over time. For instance, according to some solutions, the rest masses of quarks, electrons, and other subatomic particles began as zero some time in the very distant past and have been growing ever since. If we identify such an instant as the "creation moment," we find a natural way of describing the origin of mass. Instead of emerging all at once in the Big Bang, mass would accrue dynamically over the eons, starting at time zero. Conceivably, an epoch in which all the masses in the cosmos were somehow negative preceded that moment, which would render the creation moment merely a transition between eras—from negative to positive mass—rather than an abrupt singularity. In other words, the bang would become a blur.

Whether or not the effects of changing mass could be detected through experimentation depends on the rate of growth. The slower the increase, the less noticeable the effect. Therefore, in the limit in which mass changes by an infinitesimal amount, the five-dimensional theory would reproduce known results for standard four-dimensional relativity. There would be no detectable difference between the two theories. If, say, the rate of increase were approximately seven-billionth of a percent per year, the effect still couldn't be measured in the laboratory. Nevertheless, over the 13.7-billion-year history of the observable universe, it would amount to an increase of 100 percent in each particle—that is, from masses of zero to their current values. This dramatic increase could potentially be detected through astronomical measures.

The idea of changing mass offers an intriguing solution to several of the conceptual problems that plague conventional cosmology. Mass is not created in a sudden "big bang" singularity. Rather, it grows naturally with time, much like the familiar Hubble expansion. To prove this conjecture, however, would require new and delicate tests.

INSTANTANEOUS TIME

If the solidity of mass is a phantom, a consequence of the viewing of five-dimensional geometry through four-dimensional spectacles, could the passage of time be an illusion as well? A number of thinkers, including Fred Hoyle, J. G. Ballard, Arthur Eddington, Julian Barbour, David Deutsch, and even Einstein, have suggested that time as we know it is purely an ordering device and that the real universe is in some fundamental sense timeless.

In his science fiction novel, *October the First Is Too Late*, Hoyle offered a fictional account of some of his serious perspectives on the illusory nature of time. "There's one thing quite certain in this business," he wrote. "The idea of time as a steady progression from past to future is wrong. I know very well we feel this way about it subjectively. But we're all victims of a confidence trick. If there's one thing we can be sure about in physics, it is that all times exist with equal reality." J. G. Ballard, the well-known science fiction writer, echoed this view. In his short story "Myths of the Near Future," a character suggests that we should "think of the universe as a simultaneous structure. Everything that's ever happened, all the events that *will* ever happen, are taking place together. . . . Our sense of our own identity, the stream of things going on around us, are a kind of optical illusion."

Eddington proposed that time was subjective, a construct of the human mind. "General scientific considerations," he wrote, "favour the view that our feeling of the going on of time is a sensory impression; that is to say, it is as closely connected with stimuli from the physical world as the sensation of light is. Just as certain physical disturbances entering the brain cells via the optic nerves occasion the sensation of light, so a change of entropy. . . occasions the sensation of time succession, the moment of greater entropy being *felt* to be the later."

Oxford physicists Julian Barbour and David Deutsch have independently developed models in which each instant of time (in Barbour's terminology, "Nows") represents its own reality—a separate world, so to speak. These Nows are linked up through records of what we call the past. Thus, the only reason we say that one moment is later than another is because the "later time" contains particular information about the "earlier time." This is analogous to a film, in which each frame comprises a separate photograph. Nevertheless, if the movie is coherent, then even if these frames were cut up and placed randomly in a box, one could sort out the order of the segments.

Einstein offered his most revealing statement about the subject upon the death of his lifelong friend Michele Besso. "For us believing physicists," he said, "the distinction between past, present and future is only an illusion, even if a stubborn one."

Yet even if time has ethereal qualities, leading many prominent thinkers to question its reality, no one could doubt its driving influence in human lives. Our ghostly taskmaster cracks its whip at every junction, forcing us ever forward toward old age and death. In his later years, as poor health took its toll, Einstein was well aware of this omnipresent tormenter yet stoically managed to channel his energies toward trying to develop a unified field theory. Perhaps he believed that the mathematical elegance of the universal equations would, in the scheme of things, outweigh the petty struggles of human existence. He fervently hoped to discover a timeless model of all reality, one that could describe all forces for eternity.

Einstein's hopes for a timeless "theory of everything" never came to pass for a variety of reasons, among which was his disregard for advances in nuclear and atomic physics. Like Kaluza and Klein in their 1920s papers, Einstein failed to incorporate the strong and weak nuclear forces into his unified models. With regard to quantum theory, Einstein refused to believe that random action could pervade the process of physical observation. Moreover, he found it absurd that observers could cause the collapse of wave functions (mathematical entities in quantum mechanics containing information about particles) from a mixed system (a distribution of possible positions, for example) to a particular state (a definite location). Such interactions break the chain of determinism and assign a direction to time. The universe takes on a different character than it had in the past, merely through the actions of a single observer. Until his dying days, Einstein refused to accept a cosmos steered by capriciousness.

The collapse of quantum wave functions represents just one of many "arrows of time" in physics. Another arrow is the direction of entropy increase—the tendency for natural processes to operate in the direction of increasing disorder. Third, on the grandest scale, it's clear that the universe expands—and as recent results suggest, this expansion is eternal. It offers yet another critical distinction between past, present, and future. A fourth arrow, the direction of thought processes, may well be related to the first three—as suggested by Hawking, Penrose, and others with various models of conscious awareness. Some proof of such a connection between conscious thought and unidirectional physical processes would bolster the arguments of those who purport that all such arrows are illusions.

Communication provides still another way of characterizing time. We send signals into the future but not the past. Lighthouses brighten passing vessels with their beams only after their beacons flash. One would be astonished if a ship became illuminated before the beacon was turned on. We can employ precision instrumentation to show that there must always be a delay between the time the beam leaves the lighthouse and when it touches the ship. The directionality of this lag would provide a signpost toward the future.

All this, however, is from an external observer's outlook. Suppose someone could actually ride on the beam and determine its time of flight. (In real life, of course, such a speed-of-light journey would be impossible—but let's imagine one for the sake of argument.) According to special relativity, the time you'd experience would be the beam's proper time. Light's proper time is identical to its space-time interval—the shortest distance between two space-time events.

As we discussed in Chapter 2, relativists define space-time intervals through a variation of the Pythagorean theorem: forming the sum of the squares of the spatial distances, then subtracting the square of the time difference. Performing such a calculation with light, we arrive at the quantity zero. In other words, according to our light-speed perspective, no time would have passed at all. Therefore, you might well conclude from your "bronco ride" that time is *instantaneous*—that there is no real past or future. From your point of view, everything would have happened at once.

When it comes to general relativity, things get even stranger. As Gödel pointed out in his rotating-universe model, this theory has solutions that violate the law of cause and effect. In the proper framework, an event could even represent its own cause. If an astronaut carefully whizzed around the axis of a rotating universe, he could potentially go back in time and offer himself directions on how best to take the spin. In that case, what would be the future and what would be the past? Moreover, Thorne and his colleagues have shown that traversable wormholes (interstellar connections) could potentially serve as time machines. He, Visser, and others have developed blueprints for such hypothetical devices. In light of such danger to causality's tender threads, Hawking has proposed the "Chronology Protection Conjecture."

Could the uniqueness of history someday become an endangered concept? If the arrow of time is an artifice, would our minds be solid enough to cope with the alternative? Or like Billy Pilgrim, in Kurt Vonnegut's classic novel *Slaughterhouse Five*, would we become unhinged?

Time is complicated enough when partnered with space in a four-dimensional amalgam. Stirring extra dimensions into the mix (such as in the case of Kaluza-Klein theories) produces yet odder concoctions. Even with one extra dimension, we get strange new results, depending on the metric. Now, the metric is the specific form of the space-time interval equation, and its signature tells us how many of the terms are added (positive) and how many are subtracted (negative). Generally, positive corresponds to spacelike and negative to timelike. In a five-dimensional theory, we can in principle choose either sign for the extra part.

Most researchers assume that the extra dimension is spacelike. In that case, not only do photons travel on paths with zero interval (or separation in five dimensions), but so do massive particles like protons, or even large objects like Earth, which implies that, in some sense, all of the objects in a five-dimensional universe are in causal contact. That would explain the astonishing uniformity of the universe, without even resorting to inflation. Because communication between any two regions of the cosmos would potentially be immediate, this would also bear on the search for extraterrestrial intelligence.

However, a few researchers assume that the extra dimension of these extended theories is not spacelike but timelike. Then the results would be even weirder. Things could move faster than light, and the path of a single particle could weave in and out of space-time, like a sewing needle threading a piece of cloth. In the latter case, we could attribute the uniformity of the universe to the fact that it would consist of only a single particle—of which copies would appear everywhere as time progressed. This truly would be a grand illusion!

We see that the statement "time is instantaneous" does carry some meaning. However, perhaps a more precise characterization would be "events in the universe happen at zero interval" or "the world is simultaneous." If you doubt this statement, we're sure that Einstein, Buddha, and other "contemporaneous" figures would be up for a grand debate.

WHITHER CONSTANCY?

Changing fundamental constants, matter from higher dimensions, simultaneous time, and hidden cosmic reaches—where is this world heading, anyway? Whatever happened to the simpler days of ruler and compass, when anything you needed for measurement could be found at the local hardware store?

Yes, surveyors' tools are still for sale. Two-dimensional maps will still do just fine for taking road trips. And, don't worry, massive invisible solitons aren't invading your local swimming pool as of yet. Mundane instruments work just fine for mundane tasks. Within the bounds of our middling planet and its relatively low speed and weak gravitation, Newton remains king. However, in two directions—the extraordinarily large and the extremely small—we have long stepped beyond Sir Isaac's domain. There we turn to the help of Einstein, Bohr, Feynman, Gamow, and others but find there is still much left unexplained.

Our minds are incredible tools for pattern recognition. They have phenomenal abilities to fill in the gaps, even when accessible information is frustratingly sparse. It is wondrous, for example, how we are able to take a fact about Earth—its darkness at night, its relationship to the Moon, or its dearth of extraterrestrial signals—and extrapolate to sweeping conclusions about the cosmos itself.

Yet we certainly must recognize that the same cognitive abilities that aid us in mapping out the greater realms also have the ability to manufacture "truths" that lack validity. A good example of this statement is Dirac's Large Numbers Hypothesis. Dirac discovered what he thought was an unmistakable link between the large and the small. Nevertheless, as many present-day thinkers have concluded, his purported connection could well contain no more substance than a mirage in the desert. Or consider Kepler, who once believed that the orbits of planets were proscribed by the shapes of the regular Platonic solids (the tetrahedron, the cube, etc.). The logic seemed irrefutable and the mathematics brilliant. Nevertheless, through painstaking analysis of astronomical data he came to realize that he was wrong. His clever geometric mind had played a trick on him.

The role of thought in the universe was a dominant theme of the work of Eddington. His view of physics presents an important lesson as we press out farther and farther in our search for universal truths. The world is objective, he argued, but the *means* by which it is described, including labels such as time and space, are subjective. Hence, it would not be surprising if concepts such as heaviness, solidity, durability, and other perceptual characterizations turn out to be phantoms—important on Earth as we conduct our daily affairs but not essential to the cosmos.

Conclusion:
The Spirit of Eddington

To put the conclusion crudely—the stuff of the world is mind-stuff.

Arthur Eddington (*The Nature of the Physical World*)

THE THINKING UNIVERSE

The idea that our view of the world is invented and not discovered has a long history in philosophy. To most physicists, however, it is anathema. When Eddington in the 1930s and 1940s championed the view that science is subjective, his peers roundly lambasted him. Only a few other British scientists, who eschewed physics for metaphysics, such as A. Whitehead and E. A. Milne, took a similar stance. But recently a number of scholars have reexamined Eddington's legacy and marveled at his intellectual fortitude.

Eddington was a remarkable figure in science. In the 1920s he was one of the half-dozen people in the world who properly understood Einstein's theory of general relativity. This was a time when the competing schema of quantum theory was advancing rapidly. Appreciating both theories, Eddington tried to reconcile these starkly different worldviews. Into the mix he inserted potent Quaker beliefs that remained a vital part of his being until his premature death. Although his papers were posthumously examined and clarified by physicists, it is only now that philosophers have also come to recognize his innovative way of thinking.

Often emphasizing that he was not a solipsist, Eddington clearly stated that he believed in the existence of an external world. Nevertheless, he was convinced that our way of viewing it is limited by our biology. This conviction led him to the conclusion that science is, at least partly, subjective. His most memorable defense of this unpopular view was an analogy involving the meshsize of a fishnet.

Eddington imagined an ichthyologist investigating ocean life. He casts a net, with gaps two inches wide, into the water. When he retrieves his catch, he finds it full of fish, each more than two inches long. This leads him to generalize that no sea creature is smaller than two inches. By analogy, we retrieve from the sea of knowledge only what the mesh of our methodology allows. Other (smaller) things pass through. As Eddington pointed out, scientists are often boxed in by the boundaries of physical observation. They tend to discount what they can't directly perceive. Eddington emphasized this view when, continuing the tale of the ichthyologist, he related how difficult it can be to challenge improper scientific assumptions:

> An onlooker may object that the first generalization is wrong. "There are plenty of sea creatures under two inches long, only your net is not adapted to catch them." The ichthyologist dismisses this objection contemptuously. "Anything uncatchable by my net is *ipso facto* outside the scope of ichthyological knowledge and is not part of the kingdom of fishes. . . . In short, what my net can't catch isn't fish." Or—to translate the analogy—"If you are not simply guessing, you are claiming a knowledge of the physical universe discovered in some other way than by the methods of physical science, and admittedly unverifiable by such methods. You are a metaphysician. Bah!

In Eddington's day, labels for various physical phenomena were starting to break down. As the Copenhagen interpretation of

quantum mechanics emphasized, all particles have wavelike properties, and all forms of radiation have corpuscular features. In some
contexts, momentum is a property determined by an object's mass
and velocity. In others it has to do with an entity's wavelength.
Whether to use one or the other depends on what type of measurement an observer is making.

Bohr called this union of opposites "complementarity." Others
might call it Zen Buddhism. The phraseology of quantum physics
bears striking resemblance to the parables known as Zen koans. If a
particle crosses a detector but no one bothers to measure its velocity,
does it have a definite speed? It doesn't; quantum theory informs us.
Rather, its speed is a mixture of possibilities. How would the particle
react if placed in a magnetic field? Once again, we don't know until
the actual measurement is taken. The instant the magnet was
switched on, the wave function representing the particle would
"collapse" into one of a range of possibilities—like a jostled house of
cards falling to either the left or the right.

Like quantum physics, relativity also involves embracing seemingly
contradictory views. For example, under certain circumstances mass is
considered a feature of a solid object. In other cases it represents a pool
of energy. Once again, an observer's interaction with a body (particularly his or her relative speed) decides how much of its mass comprises
its traditional bulk and how much stems from its dynamics.

Eddington was one of the first to recognize the morphing defini-
tions of modern science. Early on he emphasized the observer's role
in any measurement. He cautioned that scientific inquiry often tells
us more about ourselves than about an "objective" external universe.
As he once summarized contemporary scientific inquiry:

> We have found a strange footprint on the shores of the unknown.
> We have devised profound theories, one after another, to account for
> its origin. At last, we have succeeded in reconstructing the creature
> that made the footprint. And lo! It is our own.

The Dynamic Trio

Uniting quantum physics and general relativity, which is one of the foremost goals of modern theoretical physics, offers a true test of embracing Eddington's call for flexibility. In some sense such a unification would involve reconciling three completely different universes. First, there is the mundane world of our immediate experiences. In this realm, time flows ever forward. Objects move, over time, along discernable paths from one position to another. For all intents and purposes, Newtonian physics works well to describe this realm.

Second, there is the Einsteinian domain, where space and time are inseparable twins. With space-time a unitary four-dimensional block, motion has a much different character. Everything, in some sense, happens at once. Like Billy Pilgrim's frozen timeline, the past, present, and future are one and the same.

Finally, there is the nebulous dominion of quantum mechanics. Its dynamics operate within a realm called Hilbert space, which, strangely enough, possesses an infinite number of dimensions. Particles don't move in this space directly. Rather, their interactions are represented through the comings and goings of wave functions. Thus, Hilbert space represents a shadow venue—not the stage for the actual drama.

Despite profound differences, these various kingdoms are closely entwined. As Einstein showed, along with his assistants Leopold Infeld and Banesh Hoffmann, general relativity can be used to derive the ordinary movements of particles, thus cementing the connection between Einstein's eternal cosmic script and the moving pen of familiar classical mechanics.

Classical mechanics is also connected to the quantum world, through the apparatus of quantum measurement. A mathematical procedure can be used to extract information from wave functions about positions, momenta, or other physical observables—though emphatically, following Heisenberg, not all of these quantities at once. Rather, if some of these properties are ascertained closely, others are known only approximately. Thus, the mundane world is a hazy, incomplete image of the mechanisms of Hilbert space.

Among this trio of physical realms, the connection between general relativity and quantum mechanics is undoubtedly the most tenuous. Physicists have been trying to solidify this link since the 1920s. The first attempt—Klein's innovative contribution to Kaluza-Klein theory—was a creative way of framing the inexactitudes of quantum physics in terms of projections from a more complete five-dimensional world. Klein's colleagues in Copenhagen admonished him, however, that he had not reckoned with the infinity of Hilbert space dimensions, only the dimensions of space-time (with one more added). Gradually, Klein and others came to realize that a full description of quantum physics needed more breathing space than standard general relativity, even extended to five dimensions, would permit.

A MATTER OF SEMANTICS

Modern physical theories have come to include both external dimensions (space, time, and any added directions) and internal dimensions that cannot be directly perceived. Such internal dimensions explain particle properties as symmetries of abstract manifolds. For example, protons can be transformed into neutrons through rotations in so-called isotopic spin space. Supersymmetry represents a similar means of rotating bosons into fermions (and vice versa) along an abstract direction. Most recently, M-theory has utilized this principle, within an 11-dimensional framework, to forge "dualities" (transformative connections) between strings and membranes of various types.

It may appear that the difference between abstract internal dimensions and physical external dimensions is purely semantic, especially in scenarios where some of the latter dimensions cannot be fully observed. Indeed, if we interpret nature through an Eddingtonian perspective, that is precisely the point. Through the power of language and reasoning, we divide the world into various categories, deeming some qualities actual and others intangible.

Take, for example, gravitation. What could feel more substantial than the pounding of rain on your head during a sudden storm or the wrenching of your body during a steep roller-coaster plunge? Thus, following the Newtonian tradition, we tend to think of gravity as a force. Even after Einstein informed us that gravity is just a curving of space-time—resulting from the patching together of various local coordinate systems—it's hard to shake the old view. A doctor would well understand a patient who complained that sudden jolts made him nauseous and might suggest a brand of motion sickness medicine. If, on the other hand, he whined that rapid coordinate system transformations made him sick, the physician might look at him askance and prescribe a very different kind of medication!

Imagine a being with no capacity for feeling physical forces, for seeing light, for hearing sound, and so on. Suppose this sensory deprivation were balanced, though, with a keen capability of perceiving geometric changes in the fabric of the universe. Not only could this being discern ripples in ordinary space-time, she could also fathom the nuances of higher dimensions. She could even sense transformations of wave functions in internal dimensions. What could she tell us about the cosmos?

If there were some way of communicating with such a being, we would learn that geometry is real and that mass, force, time, and so forth are all illusions. Our attempts to describe perceptions such as heaviness, loudness, darkness—in short, everything familiar—would likely be met by sheer disbelief. Even the chronicles of our lives, laid out over time, would have absolutely no meaning to such a timeless creature. In Eddington's parlance, our worlds would be as different as our minds.

As mortals with mental and sensory limitations, we cannot know everything about the cosmos. The mathematician Kurt Gödel made this point when he demonstrated that no logical system is complete. His findings clashed sharply with attempts by Hilbert to systematize the mathematical and physical universes. Perhaps this is for the best. As the famed Argentine writer Jorge Luis Borges described in such short stories as "The Aleph" and "The Zahir," a person to whom every facet of reality was suddenly revealed could well go mad.

Maybe it is fortunate, then, that our knowledge of the universe arrives by dribs and drabs. Photon by photon, we slowly drink in one particular type of cosmic energy, leaving us ample time to savor (and interpret) this brew. Through gravitational detectors, we soon hope to savor a different type of broth. As connoisseurs of such stellar ferment, we pride ourselves in our growing appreciation of what we sip. Yet we also realize that much lies in the bottom of the barrel, inaccessible to our tasting, and that all this could have quite an alien flavor.

Truth and Mathematics

Eddington wrestled with the question of how best to interpret the limited information about the universe revealed to us through science. This issue remained of paramount importance to him throughout his life. If the bulk of the cosmos is composed of dark matter, dark energy, and perhaps even inaccessible extra dimensions, how can we best extend our current understanding to plumb at least part of these hidden depths? For example, what physical labels should we assign to these higher dimensions? Is there a unique mapping between observational data and physical actuality, or could the truth be a hydra of countless faces?

As Eddington emphasized, there are as many ways to describe our world as there are intelligent observers. Each cognizant being interacts with reality uniquely. There needs, however, to be an interaction. In contrast to Kant, Eddington was doubtful that pure reasoning could lead to new scientific knowledge.

Eddington's personal means of understanding, especially in his later years, centered on the mathematical subjects of number theory and tensor calculus. He was obsessed with finding simple numerical patterns that would encompass physical truths. His preference for mathematical methods stemmed, no doubt, from his own exceptional skills in that field. However, he made it clear in his writings that he was open to different approaches (including the religious one involving the Friends' Meeting House, which he attended regularly). He chose mathematics because it seemed to him to be the most *effective* means of description.

This view, while disputable, is nevertheless pragmatic. It is indeed this view that underlies much of modern work in the physics of fields and particles. The present emphasis on descriptions of the world in terms of higher-dimensional geometry is analogous to the extension of ordinary two-dimensional chess to the three-dimensional variety now available; both represent a trend to increased sophistication. But as in the case of 3D chess, physics in higher dimensions needs to *invent* new rules of play. The nature of such rules will likely spark debate for quite some time.

Reduction of the mechanical concepts of physics to the more intuitive ones of philosophy is an ongoing process. Eddington, in recognizing this, left a major legacy for both fields.

ALL THE MYRIAD WAYS

It is strange to think that the truths we discover about our universe may not be true for all possible universes. Thus, even if a modern Eddington stumbled on mathematical relationships that precisely define the space we see, there is no guarantee that all possible realms would have the same relationships. Maybe other universes exist with three types of electrical charge, dozens of fundamental forces, and thousands of stable chemical elements. Perhaps, somewhere in the labyrinth of possibilities, there are places where time runs backward and black holes light up the skies.

In the many-worlds interpretation of quantum mechanics, proposed by Hugh Everett and championed by Bryce DeWitt, each time a subatomic process has several possible results, the universe bifurcates. Each copy is identical save for one distinction—the particular quantum process in question has a different outcome. For example, in one version a certain electron might jump from one atomic level to another, triggering the release of a photon, while in all others no such transition occurs. After bifurcation each parallel realm carries on as if nothing had happened. No trace exists of the alternatives, save in the minds of speculative thinkers.

Many modern cosmological models similarly embrace the notion of parallel universes: alternative realities that coexist side by side with our own. Chaotic inflation, the self-reproducing universe, and kindred descriptions of the cosmos conceive of a gardenlike multiverse sprouting various types of plants—some gentle, some quite prickly. Because the cosmic horizon's high picket fence would hide such exotic growth from our view, testing such scenarios would be challenging. The limits of luminous communication are daunting indeed. Imagine what fantastic possibilities would await, however, if we could somehow jump this fence and explore other patches.

In Larry Niven's classic short story, "All the Myriad Ways," a future corporation—called Crosstime—develops the means for contact between alternative realities, offering people access to all the worlds that could have been. Each possibility, no matter how strange, has its own romping ground wherein its events could be played out. As Niven describes this jumble:

> There were timelines branching and branching, a megauniverse of universes, millions more every minute. . . . The universe split every

time someone made a decision. . . . Every choice made by every man, woman and child on Earth was reversed in the universe next door. . . .

At least one innovative young thinker, MIT physicist Max Tegmark, believes that parallel-universe models are "empirically testable, predictable and falsifiable." Originally from Stockholm, Tegmark arrived at his present position by way of Princeton, where he had the opportunity to collaborate with John Wheeler. Like Wheeler he enjoys pondering the most far-flung interpretations of physics, while simultaneously conducting more mainstream research. Tegmark refers to the former as "crazy stuff."

At a Princeton conference held in 2002 in honor of Wheeler's 90th birthday, various physicists and other experts seemed engaged in a contest to paint the most all-encompassing portraits of the cosmos. In terms of far-reaching schemes, Tegmark arguably outdid his colleagues, however, by framing their proposals and others in terms of an intricate multitiered labyrinth of parallel realms. He asserted that the infinite expanse of space made it certain that there exist multiple copies of every person, place, and thing in the cosmos. These replicas cannot be observed because they lie well beyond the Hubble radius.

"Is there another copy of you reading this," asked Tegmark in an article summarizing his talk, "deciding to put it aside without finishing this sentence while you are reading on? A person living on this planet called Earth, with misty mountains, fertile fields and sprawling cities, in a solar system with eight other planets. The life of this person has been identical to yours in every respect—until now, that is when your decision to read on signals that your two lives are diverging."

Tegmark classified parallel universes into four distinct categories. Level One, he suggested, included parts of space that lie forever outside the range of telescopes. They would be "parallel" in the sense that they'd contain many duplicates and near duplicates, arising through the vagaries of chance. This is a spatial version of Nietzsche's old idea of "eternal return": given a finite number of atomic configurations and an unlimited amount of time, random actions would eventually produce clones. Somewhere, perhaps, a duplicate of Nietzsche is writing a paper pointing out that infinite space could serve just as well as infinite time in reproducing all structures again and again. And maybe in another remote corner of reality a third Nietzsche is preparing to sue the second for plagiarism.

As strange as other Level One realms would seem, at least they'd have the same natural laws as ours. In Level Two regions, on the other hand, the fundamental constants, properties of elementary particles, and so forth could well be very different. The second tier in Tegmark's scheme comprises the set of all "bubble universes" produced during chaotic inflation. Because each bubble would begin its life as a simple quantum fluctuation with few of its attributes set, it could develop in multifarious ways. Successful bubbles would grow infinitely large, potentially evolving into viable universes. Others would fizzle out, faltering before they had a chance to generate structures. They'd vanish back into the primordial nothingness.

Blessed with ample time, the bubbles that did happen to grow would pass through numerous stages of symmetry breaking. During each transition, unified fields would break down into various interactions, and simple particles would give birth to complex menageries with assorted masses and properties. Depending on special models of development, some of these steps could coincide with certain dimensions curling up. Alternatively, all of the dimensions could remain of equal magnitude.

The timing and order of these phases would be specific to each bubble, resulting in diverse possibilities for the strengths of different forces and the masses and types of different particles. So, for instance, in some bubbles the charges of the proton and electron would be very different, precluding the formation of neutral atoms. Naturally, such conditions would hinder the growth of stable structures.

As each universe matured, it would generate regions of strong gravity. These would serve as spawning grounds to produce new fluctuations. These fluctuations, in turn, could evolve into new bubble universes—generating more and more generations. One might wonder how a multiverse could accommodate so many bubble universes, each of unlimited size. Fortunately, infinity's hotel always has room for new guests.

A variation of this bubble geneology is Perimeter Institute physicist Lee Smolin's notion of cosmic survival of the fittest. Smolin has constructed a clever biological analogy between the replication of universes and the reproduction of living organisms. Black holes, he has asserted, would offer ideal wombs for the gestation of baby universes. Therefore, universes with more black holes could produce more offspring and tend to dominate over competing cosmologies.

Each time a baby universe emerged, its fundamental constants would be different—the equivalent of genetic mutation. Some of the changes would auger well for nucleosynthesis, producing massive stars that would eventually collapse into black holes. Other alterations would turn out to be duds—allowing few or no stars to reach maturity. Naturally, these would have far fewer black holes—and less opportunity to breed. Their "genetic" lines would thus tend to die out over time.

Now here's the clincher—because universes favorable to the production of many black holes would be well suited for nucleosynthesis, they would also produce many vibrant stars like the Sun, well suited for habitable planets. Hence, these universes would also tend to have the conditions favorable to support life. The process of natural selection would thereby explain the emergence of living beings, justifying why conditions in the cosmos are so supportive.

In addition to eternal inflation scenarios and Smolin's evolutionary idea, Tegnark also grouped brane-world models into Level Two. However, he pointed out that other branes would gravitationally interact with ours, rendering them more symbiotic than separate. Therefore, their status as truly parallel would not be quite as solid.

Next in Tegmark's scheme comes Level Three, his designation for Everett's many-worlds interpretation. This level has an altogether different character than the first two, since it is quantum mechanical—not cosmological—in origin. If this model were correct, reality would bifurcate each time an experimenter made a subatomic measurement. Therefore, unlike the other possibilities for alternative worlds, the production of parallel realities would transpire right here and now.

Finally, Level Four, the most abstract grouping of all. It includes the set of all conceivable mathematical structures. A mathematical structure is an axiomatic system in which certain suppositions imply a variety of theorems. Euclidean and non-Euclidean geometries represent examples of these. We can show that there is an unlimited range of possible mathematical rules that would produce a never-ending assortment of relationships. For instance, in some realities there would be five Platonic solids (regular polyhedra such as cubes), in others there would be 10, and in yet others there would be an infinite number.

Why should varying mathematical relationships make a difference to the material universe? Just as flat universes and curved universes, because of their differing geometries, have distinct physical properties, any novel axiomatic system engenders a new physics. Hence, unlimited types of mathematical structure would correspond to a plethora of divergent realities. Tegmark called this state of affairs "mathematical democracy."

Though highly speculative, Tegmark's talk was one of the many highlights of a truly thought-provoking conference. Other talks included DeWitt describing the many-worlds hypothesis, Randall discussing warped dimensions, Smolin delving into quantum gravity, Linde speaking about inflation, and so forth. All the while, a gratified Wheeler sat at the front of the hall, sampling the philosophical fruit of the many gardens of inquiry he had nurtured.

By sheer coincidence, shortly after the Wheeler commemoration, Tegmark and one of us (Halpern) found ourselves members of the "jury" at a production of Michael Frayn's acclaimed play

"Copenhagen." The play is about the changing relationship of Bohr and Heisenberg before and during the Second World War, when they found themselves working for opposite sides. (Heisenberg was involved in the Nazi nuclear fission program—but the court of history has not rendered a verdict on whether he helped or hindered it.) One of the themes of the drama is that quantum uncertainty allows for simultaneous alternative realities—such as Bohr and Heisenberg being friends (because of their long-standing collaboration) and foes (because of the war) at the same time. At least in some productions a few members of the audience are seated in a "jury box" on stage—presumably to render a verdict on Heisenberg's intentions. Thus, we found ourselves on the same panel, watching and judging the show. Tegmark appeared to enjoy seeing these alternative realities play out—like parallel realms in the multiverse of history.

Indeed, confined to our small enclave of space, we are all jurors, rendering a verdict on the unfolding cosmic drama. Like any jury, our varied prejudices and perspectives affect the outcome of our conclusions. Each of us decides what seems to be "crazy stuff" and what appears to be mainstream.

No measurement we make is wholly independent of our human experiences. Because we filter all information through our perceptions, in some sense we generate our own parallel universes—each a different facet of a multifarious prism. Hence, as Eddington pointed out, even if there is a true reality, it could well be lost in the mirror maze of subjectivity.

Triumph and Its Aftermath

From the lowly vantage point of Earth, our instruments and intuitions have propelled us billions of light-years into the void and eons back in time. Questions unanswered for millennia have finally found credible answers. The ancient philosopher's quest for the age of the heavens has in some way been resolved, with the knowledge that 13.7 billion years have passed since the primordial fireball let loose its power. Like exacting surveyors, we have scoped out the shape of visible space. In a wry twist on the legacy of Columbus, we can finally proclaim that the universe is flat—at least in three dimensions and possibly in five dimensions. Cosmology has ample reason to glow in triumph.

In the particle realm, scientists similarly have much cause for celebration. Two of the four forces of nature are united as the electroweak theory, a highly successful physical model with astonishing predictive powers. As for the strong interaction, quantum chromodynamics remains widely accepted. It is more difficult to work with than the other models but nevertheless seems to serve well. With regard to gravity, true, there's no quantum theory as of yet. But at least it is well described by Einstein's remarkable theory. So far, all known measurements of general relativity appear to verify its validity. Optimism abounds in the superstring community that a "theory of everything" will soon be forthcoming.

Many times in the history of knowledge, various thinkers have proclaimed the imminent end of science. Practically all there is to know, they've asserted at such moments, has already been discovered. For example, in the late 19th century, physicists considered Newtonian physics a perfect description of mechanics and Maxwell's equations a complete model of electromagnetism and light. Though these theories harbored mutual contradictions, many scientists believed that the existence of aether could help explain these. Physics seemed virtually complete. Only a few "minor mysteries," such as the reason for discrete spectral lines and the origins of radioactivity (discovered in 1895), appeared to remain. Nevertheless, it was those very conundrums that opened up the floodgates, ushering in waves of new scientific activity.

Today cosmology has arrived at a concordance model—one that meshes well with all known data. Probes of distant supernovas, gravitational lensing measurements, and readings of the cosmic microwave background have pinned down cosmological parameters to an unprecedented level. Yet what is so striking about the new results is that science once again faces gaps and contradictions. So much of the substance and power of the cosmos simply cannot be explained. Addressing these hidden materials and forces could well spark a revolution in physics as far-reaching as that of the early 20th century.

As we have seen, theorists have been off to a good start. From models with changing mass to those with variable speed of light, and from various recipes for inflation to assorted prescriptions for higher-dimension dynamics, there seems no end to clever ideas for resolving the deepest mysteries of the cosmos. One common theme is that the simplest form of general relativity could require some type of modification—be it by simply restoring the cosmological constant, adding additional fields, or extending it through extra dimensions.

Some of these novel schemes, however, explain what we can or cannot observe by positing vast new sectors of reality—parallel universes, of various sorts. This can be a tricky business. By positing new territories that could never be explored, we render a theory essentially nontestable. The best new models have clear predictions that allow for careful matching with experiment data. "Observation," as Eddington once wrote, "is the supreme Court of Appeal."

One of the greatest mysteries arises when we turn to the cosmic future. Current scenarios suggest that the universe will expand forever. Some researchers have attempted to map out the far future of the universe, painting a bleak picture of the slow demise of all vibrant entities—from stars to life. Like the grim reaper, entropy will eventually cloak the cosmos in absolute darkness. Even more terrifying is the possibility of a "big rip"—the tearing apart of the fabric of the universe: the ultimate doomsday.

If scenarios for cosmic demise remind us of Western apocalyptic notions, the alternative is reminiscent of Eastern views of endless renewal. In oscillatory scenarios, such as the cyclic model, the universe will eventually be reenergized. Heeding Eddington's words, it will be interesting to see what evidence accrues for each of these possibilities.

THE FUTURE OF COSMOLOGY

Luckily, researchers are planning a number of exciting new experiments that will help sort the theoretical wheat from the chaff. Due in part to unfortunate budget cuts in American experimental programs, the center of activity for fundamental science has largely shifted to Europe. Therefore a number of the planned experiments will take place under the auspices of CERN and the ESA.

CERN's flagship project, the Large Hadron Collider (LHC), will be the most powerful particle accelerator in the world. Scheduled to begin operations in 2007, it will have the capability of smashing together beams of protons at energies of 14 TeV (approximately two-millionths of a Joule). Although a fraction of a Joule (much smaller than a nutritional calorie) may not seem like much, that is significantly higher than the capacity of its closest contender, the Tevatron at Fermilab. Moreover, these energies are concentrated in an incredibly tiny region of space.

Experiments designed for the LHC include searches for supersymmetric companion particles, a hunt for the Higgs boson (an essential missing ingredient of modern field theories, believed to have an ultrahigh mass), and tests to discern if gravitons vanish from certain collisions (and presumably escape into a higher dimension). Given that modern cosmology has many ties to particle physics, these experiments would help distinguish various models of the universe. For example, if experimenters find that certain byproducts of a collision are missing, suggesting that the gravitons produced in the crash have escaped into another dimension, this result would offer a boost for scenarios based on large extra dimensions.

The long-awaited year of LHC's inauguration coincides with the launching of a major space probe, the Planck satellite. Sponsored by the ESA, it represents the next step beyond WMAP for precise measurements of minute anisotropies in the cosmic microwave background. Its intended final orbit, approximately 1 million miles away, offers an ideal situation for taking sensitive temperature readings far from the influences of the Earth, Moon, and Sun.

The Planck satellite's great precision will enable it to discern tiny changes in the fine-structure constant—one part in a thousand over the age of the universe. This accuracy will substantially improve on the bounds set by WMAP and other instruments. Pinning down whether or not alpha varies has wide-ranging implications, given that many higher-dimensional theores predict a small change over time. Thus, it will be riveting to see on what side the Planck data come down—variation or not.

Another important gauge of whether or not the natural constants are changing involves another satellite, Gaia. Originally an acronym for the Global Astrometric Interferometer for Astrophysics but later modified, Gaia is scheduled for launch by the ESA in 2011. This probe will be the most precise astronomical mapping device in history, pinpointing the exact distances and movements of billions of stars. Two scientific instruments placed on board will serve to collect and analyze light from large sectors of the sky. A Russian Soyuz rocket will help propel Gaia into the same orbital region occupied by Planck, granting it a similarly clear view.

Performing such detailed measurements will place Gaia in the ideal position to measure changes in the gravitational constant. The motions of interacting celestial bodies, such as binary star systems, strongly depend on the form of the law of gravity. If the constant driving that relationship has altered in any way over time, Gaia would have the capability of recording such discrepancies.

Less than a century ago, Hubble revealed the cosmos to be a vibrant structure, full of explosive energy. For the first time in history, humankind realized that from the smallest meteors to the largest galaxies the heavens were in a constant state of flux. Newton's hallowed jewel, the delicate latticework of fixed stars, seemed to fracture like shattered glass. General relativity stood as the perfect means of modeling a dynamic universe, yet its author was reluctant at first to step away from stability. Gradually he and others came to accept an evolving cosmos. Through thinkers such as Eddington, Lemaitre, and Gamow, the world came to appreciate the significance of this radical new perspective.

Today we face another revolution in cosmology. Astonishing new findings challenge explanation. Unlike the discoveries of the 1920s, no widely accepted theory accounts for all the recent results. Contending theories, such as those with a changing gravitational constant, variable light speed, quintessence, colliding membranes, extra dimensions, and so forth, call for fundamental alterations in our conception of the universe. We cannot yet tell which (if any) of these will direct researchers along the path of truth. If any of these theories pass the test of experiment, it will undoubtedly launch physics into an astonishing new era. Given our species' insatiable curiosity, we surely will not rest until the great cosmic conundrums are resolved. Until then, as the saying goes, there is joy in the journey.

Acknowledgments

Paul Halpern thanks his family and friends for their support, including his wife, Felicia, and his sons, Eli and Aden. He would also like to acknowledge the generous assistance of the John Simon Guggenheim Memorial Foundation and the University of the Sciences in Philadelphia for the opportunity to take a research sabbatical. He expresses his sincere gratitude to the Department of Physics and Astronomy at Haverford College for a visiting professorship, during which time much of his contribution to this book was written. He particularly thanks Bruce Partridge and Steve Boughn for interesting discussions about the history of modern cosmology; Fronefield Crawford for his comments about pulsar astronomy; and Suzanne Amador-Kane, Jerry Gollub, and Walter Smith. Thanks also to Francis Everitt, Rai Weiss, Paul Steinhardt, Saul Perlmutter, Raman Sundrum, Alan Chodos, Fred Cooperstock, Kumar Shwetketu Virbhadra, and Engelbert Schucking for their insightful comments. Above all, he thanks his coauthor Paul Wesson for suggesting this collaborative project.

Paul Wesson thanks Pat and Sterling, whose optimism is as unbounded as the universe. He also acknowledges the time spent at various universities over the years, where the ideas in this work were allowed to germinate (Berkeley, Cambridge, Stanford, and Waterloo). Some of this work was carried out during a stay arranged by Francis Everitt at Gravity Probe B, at Stanford. His views were enlightened by numerous colleagues, including Stuart Bowyer, James Overduin, Hongya Liu, Bahram Mashhoon, and Jaime Ponce de Leon. Ideas in cosmology often require dedicated technical work to show their viability, and in this regard he thanks his graduate students, notably Dimitri Kalligas, Tomas Liko, and Sanjeev Seahra. The inspiration for many of the topics discussed here goes back to Sir Arthur Eddington, and it is a pleasure to know that Paul Halpern shares the respect due this great thinker.

Both Pauls acknowledge the help and suggestions provided by Giles Anderson of the Anderson Literary Agency. We appreciate the comments of John Huchra and Rabindra Mohapatra in reading over a draft of the manuscript. Many thanks to our editor, Jeff Robbins, our project editor, Sally Stanfield, and the staff of the Joseph Henry Press for useful advice and assistance.

Notes

1
To See the World in a Grain of Sand

p. 12 **Your waistline may be spreading:** Richard Price, quoted in Zeeya Merali, "Why the Universe Is Expanding Without Us," *New Scientist*, October 1, 2005, p. 13.

p. 12 **Were the succession of stars endless:** Edgar Allen Poe, *Eureka: A Prose Poem* (Putnam: New York, 1848), p. 102.

p. 14 **Resolution of Olbers' paradox:** Paul S. Wesson, K. Valle, and R. Stabell, "The Extragalactic Background Light and a Definitive Resolution of Olbers' Paradox," *The Astrophysical Journal*, vol. 317, pp. 601-606 (1987).

p. 16 **If all places:** Isaac Newton, "On Universal Design," in *Newton's Philosophy of Nature*, H. S. Thayer, ed. (New York: Hafner Publishing Co., 1953), p. 67.

2
Infinity in the Palm of Your Hand

p. 28 **vicious, inflexible, quarrelsome:** Arthur Koestler, *The Sleepwalkers: A History of Man's Changing Vision of the Universe* (New York: Macmillan, 1959), p. 229.

p. 28 **Yet as I reflected:** Ibid., p. 259.

p. 33 **Absolute space:** Isaac Newton, "Fundamental Principles of Natural Philosophy," in *Newton's Philosophy of Nature*, H. S. Thayer, ed. (New York: Hafner Publishing Co., 1953), pp. 17-18.

p. 33 **Relative quantities:** Ibid., p. 23.

p. 33 **Instead of referring a moving body:** Ernst Mach, *The Science of Mechanics*, translated by Thomas McCormack (Chicago: Open Court, 1907), p. 234.

p. 34 **The fact that the needle:** Albert Einstein, "Autobiographisches," in *Albert Einstein als Philosoph und Naturforscher [Albert Einstein as a Philosopher and Scientist]*, Paul Arthur Schilpp, ed. (Braunschweig: Vieweg, 1979), p. 7. Quoted in Albrecht Fölsing, *Albert Einstein*, translated by Ewald Osers (New York: Penguin, 1997), p. 14.

p. 39 **How does it come:** Albert Einstein, *Autobiographical Notes*, translated and edited by Paul Arthur Schilpp (La Salle, Ill.: Open Court, 1979), p. 59.

p. 54 **That was my problem:** Carl Sagan, interviewed on *Nova: Time Travel.*

p. 60 **We know almost nothing:** Bryce DeWitt, reported in *The First William Fairbank Meeting on Relativistic Gravitational Experiments in Space*, M. Demianski and C. W. F. Everitt, eds. (Singapore: World Scientific, 1993), p. xvii.

p. 66 **My father:** C. W. Francis Everitt, personal communication, January 26, 2005.

p. 70 **Jerrold said to me:** Interview with Rainer Weiss, Haverford College, April 5, 2005.

p. 72 **Observing gravitational waves:** Ibid.

3
Eternity in an Hour

p. 76 **committed something in the theory:** Letter from Albert Einstein to Paul Ehrenfest, February 4, 1917. Quoted by Albrecht

Fölsing in *Albert Einstein*, translated by Ewald Osers (New York: Penguin, 1997), p. 387.

p. 81 **grumpy letter:** Edward Harrison, *Cosmology: The Science of the Universe* (Cambridge: Cambridge University Press, 1981), p. 297.

p. 82 **Not for a moment:** Ibid., p. 307.

p. 87 **It seems that this "alphabetical" article:** Letter from George Gamow to Oskar Klein, undated (probably 1948), Niels Bohr Archive.

p. 87 **Thank you very much:** Letter from Oskar Klein to George Gamow, undated (probably 1948), Niels Bohr Archive.

p. 90 A third alternative cosmology, originally proposed by Klein and later modified by Nobel laureate Hannes Alfvén, never quite caught on. It imagined the cosmos as a contracting, then exploding, plasma of matter and antimatter.

p. 92 **There is one unfortunate:** R. H. Dicke, *A Scientific Autobiography*, 1975, unpublished. Cited by W. Harper, P. J. E. Peebles, and D. T. Wilkinson in "Robert Henry Dicke: A Bibliographical Memoir," Princeton University Physics Department, unpublished manuscript, p. 6 (2001).

p. 93 **Throughout most of the history:** Steven Weinberg, *The First Three Minutes: A Modern View of the Origin of the Universe* (New York: Basic Books, 1977), p. 4.

p. 94 **I cannot deny:** Ibid., p. 9.

p. 103 **One of the things he talked about:** Alan Guth, quoted by Alan Lightman and Roberta Brawer in *Origins: The Lives and Worlds of Modern Cosmologists* (Cambridge, Mass.: Harvard University Press, 1990), pp. 470-471.

p. 104 **he was generous:** Alan Guth, *The Inflationary Universe: The Quest for a New Theory of Cosmic Origins* (Reading, Mass.: Addison Wesley, 1997), p. 207.

p. 107 **I had a telescope:** Interview with Paul Steinhardt, Princeton University, November 5, 2002.

p. 112   **The Big Bang: Dead or Alive?,** *Sky and Telescope,* May 1991, p. 467.

p. 113   **I always enjoyed:** Saul Perlmutter, "Did you ever wonder. . . what dark energy accelerates the universe?," *Lawrence Berkeley Laboratory Report, http://www.lbl.gov/wonder/perlmutter.html.*

p. 116   **I think people:** Interview with Paul Steinhardt, Princeton University, November 5, 2002.

p. 118   **Fortunately Andrei:** Michael Turner, talk given at an American Physical Society meeting, April 5, 2003.

4
DARKNESS APPARENT

p. 122   **If it is tidal debris:** Robert Minchin et al., "A Dark Hydrogen Cloud in the Virgo Cluster," Cardiff University preprint, p. 1 (2004).

p. 123   **In 1962, with my students:** Vera Rubin, "A Brief History of Dark Matter," Swarthmore College Colloquium, February 25, 2005.

p. 135   **The new upgrade:** Leslie Rosenberg, quoted by Ann Parker in "Small Particle May Answer Large Physics Questions," *Science and Technology Review,* Lawrence Livermore National Laboratory, January/February 2004, pp. 9-11.

p. 139   **string theory is a piece:** Ed Witten, "On the Right Track," *Frontline,* vol. 18, no. 3 (2001).

p. 140   **Unfortunately, the microrealm:** John Horgan, "Is Science a Victim of Its Own Success?," *APS News,* December 1996.

p. 141   The idea that invisible halo objects could be composed of mirror matter was first put forth by Rabindra N. Mohapatra and Vigdor L. Teplitz in "Mirror Matter MACHOs," *Physics Letters B,* vol. 462, p. 302 (1999).

p. 142   **believed that he might be able:** H. G. Wells, "The First Men in the Moon," *Seven Science Fiction Novels,* (New York: Dover, 1934), p. 467.

p. 143 **The old question:** Phone interview, Alan Chodos, December 6, 2002.

p. 143 **At the time:** Phone interview, Raman Sundrum, December 3, 2002.

p. 146 For an account of the evolution of Darwin's finches on Galapagos, see Jonathan Weiner, *The Beak of the Finch: A Story of Evolution in Our Time* (New York: Knopf, 1994).

5
EVER-CHANGING MOODS

p. 150 **Nobody knew him very well:** Phone interview, Engelbert Schucking, December 9, 2002.

p. 150 **It is nice, but:** George Gamow, *Thirty Years that Shook Physics* (New York: Doubleday, 1966), p. 121.

p. 156 Itzhak Goldman, "Limits of G-Variability from Spin-down of Radio Pulsars," *The First William Fairbank Meeting on Relativistic Gravitational Experiments in Space*, M. Demianski and C. W. F. Everitt, eds. (Singapore: World Scientific, 1993), pp. 9-14.

p. 160 **We cannot decide immediately:** Pascual Jordan, *The Expanding Earth: Some Consequences of Dirac's Gravitation Hypothesis*, translated and edited by Arthur Beer (New York: Pergamon Press, 1971), pp. 156-157.

p. 160 **The idea of changing G:** Phone interview, Engelbert Schucking, December 9, 2002.

p. 162 **lifelong obsession:** João Magueijo, *Faster than the Speed of Light: The Story of a Scientific Speculation* (Cambridge, Mass.: Perseus, 2003), p. 139.

p. 165 **found that there was a considerable potential:** Robert L. Oldershaw, "Self-Similar Cosmological Model: Introduction and Empirical Tests," *International Journal of Theoretical Physics*, vol. 28, no. 6, pp. 669-694 (1989).

p. 170 The slightly more complex rule relating angular momentum and mass is described by Wlodzimierz Godlowski,

Marek Szydlowski, Piotr Flin, and Monika Biernacka in "Rotation of the Universe and the Angular Momenta of Celestial Bodies," *General Relativity and Gravitation*, vol. 35, no. 5, pp. 907-913 (2003).

p. 170   A possible connection between dark energy and the rotation of the universe is explored by Wlodzimierz Godlowski and Marek Szydlowski in "Dark Energy and Global Rotation of the Universe," *General Relativity and Gravitation*, vol. 35, no. 12, pp. 2171-2187 (2003).

6
ESCAPE CLAUSE

p. 176   **the ultimate free lunch:** Alan Guth, quoted by Marcia Bartusiak in *Through a Universe Darkly: A Cosmic Tale of Ancient Ethers, Dark Matter and the Fate of the Universe* (New York: HarperCollins, 1993), p. 247.

p. 178   **To his amazement:** Abdus Salam, interviewed on "NOVA: What Einstein Never Knew," originally broadcast October 22, 1985.

p. 179   **Einstein asked a question:** Theodor Kaluza, Jr., interviewed on "NOVA: What Einstein Never Knew," originally broadcast October 22, 1985.

p. 188   **The folded universe picture:** Nima Arkani-Hamed, Savas Dimopoulos, Nemanja Kaloper, and Gia Dvali, "Manyfold Universe," *Journal of High Energy Physics*, vol. 12, p. 4 (2000).

p. 189   **When I find a model of physics:** Phone interview, Raman Sundrum, December 3, 2002.

p. 190   **I've been waiting for string theory:** Interview with Paul Steinhardt, Princeton University, November 5, 2002.

p. 192   **Once the universe:** Ibid.

7
WHAT IS REAL?

p. 196 **[General relativity] is sufficient:** Albert Einstein, "Physics and Reality," In *Ideas and Opinions,* Carl Seelig, ed., (New York: Crown Publishers, 1954), p. 311.

p. 203 **a rounded, smooth and well-defined heap:** John Scott Russell, "Report on Waves," *Report of the Fourteenth Meeting of the British Association for the Advancement of Science* (London, 1845), pp. 311-390.

p. 206 **There is no fundamental difference:** C. Brans and R. H. Dicke, "Mach's Principle and a Relativistic Theory of Gravitation," *Physical Review,* vol. 124, no. 3, p. 927 (1961).

p. 208 **There's one thing quite certain:** Fred Hoyle, *October the First Is Too Late,* (New York: Harper and Row, 1966), p. 75.

p. 208 **think of the universe:** J. G. Ballard, "Myths of the Near Future," *The Complete Short Stories of J. G. Ballard* (London: Flamingo, 2001), p. 1078.

p. 208 **General scientific considerations:** Arthur Eddington, *The Philosophy of Physical Science* (New York: Macmillan, 1939), p. 197.

p. 209 **For us believing physicists:** Letter from Albert Einstein to Vero and Bice Besso, March 21, 1955. Translated and quoted by Banesh Hoffmann, with Helen Dukas, in *Albert Einstein: Creator and Rebel* (New York: New American Library, 1972), pp. 257-258.

CONCLUSION

p. 216 **An onlooker may object:** Arthur Eddington, *The Philosophy of Physical Science,* p. 16.

p. 217   **We have found a strange footprint:** Arthur Eddington, *Space, Time and Gravitation* (Cambridge: Cambridge University Press, 1921), p. 201.

p. 223   **There were timelines branching:** Larry Niven, *All the Myriad Ways* (New York: Ballantine Books, 1971), p. 1.

p. 224   **empirically testable:** Max Tegmark, "Parallel Universes," *Science and Ultimate Reality: From Quantum to Cosmos*, J. D. Barrow, P. C. W. Davies, and C. L. Harper, eds. (Cambridge: Cambridge University Press, 2003), p. 21.

p. 224   **Is there another copy:** Ibid.

p. 230   **Observation:** Arthur Eddington, *The Philosophy of Physical Science*, p. 9.

Further Reading

(Technical works are marked with an asterisk.)

Barrow, John, *The Constants of Nature: From Alpha to Omega—the Numbers That Encode the Deepest Secrets of the Universe* (New York: Pantheon, 2003).

Bartusiak, Marcia, *Through a Universe Darkly: A Cosmic Tale of Ancient Ethers, Dark Matter and the Fate of the Universe* (New York: HarperCollins, 1993).

Bergmann, Peter, *The Riddle of Gravitation* (New York: Charles Scribner's Sons, 1968).

Croswell, Ken, *The Universe at Midnight: Observations Illuminating the Cosmos* (New York: The Free Press, 2001).

Davies, Paul, and John Gribbin, *The Matter Myth: Dramatic Discoveries That Challenge Our Understanding of Physical Reality* (New York: Simon and Schuster, 1992).

Eddington, Arthur, *Space, Time and Gravitation* (Cambridge: Cambridge University Press, 1921).

Eddington, Arthur, *The Nature of the Physical World* (Ann Arbor: University of Michigan Press, 1981).

Gibbons, Gary, "Brane-Worlds," *Science*, vol. 287, January 7, 2000, pp. 49-50.

Goldsmith, Donald, *The Runaway Universe: The Race to Discover the Future of the Cosmos* (Reading, Mass.: Perseus, 2000).

Greene, Brian, *Fabric of the Cosmos: Space, Time and the Texture of Reality* (New York: Knopf, 2004).

Gribbin, John, *The Omega Point: The Search for the Missing Mass and the Ultimate Fate of the Universe* (New York: Bantam Books, 1988).

Gribbin, John, and Paul Wesson, "Fickle Constants of Physics," *New Scientist*, vol. 135, no. 1828, p. 30 (1992).

Gribbin, John, and Paul Wesson, "The Fifth Dimension of Mass," *New Scientist*, vol. 119, no. 1631, p. 56 (1988).

Guth, Alan, *The Inflationary Universe: The Quest for a New Theory of Cosmic Origins* (Reading, Mass.: Perseus, 1998).

Halpern, Paul, *Cosmic Wormholes: The Search for Interstellar Shortcuts* (New York: Dutton, 1992).

Halpern, Paul, *The Cyclical Serpent: Prospects for an Ever-Repeating Universe* (New York: Plenum, 1995).

Halpern, Paul, *The Great Beyond: Higher Dimensions, Parallel Universes and the Extraordinary Search for a Theory of Everything* (Hoboken, N.J.: Wiley, 2004).

Halpern, Paul, *The Quest for Alien Planets: Exploring Worlds Beyond the Solar System* (New York: Plenum, 1997).

Halpern, Paul, *The Structure of the Universe* (New York: Henry Holt, 1996).

Hawking, Stephen, *The Universe in a Nutshell* (New York: Bantam, 2001).

Kaku, Michio, *Einstein's Cosmos: How Albert Einstein's Vision Transformed Our Understanding of Space and Time* (New York: W. W. Norton, 2004).

Kaku, Michio, *Parallel Worlds: A Journey Through Creation, Higher Dimensions and the Future of the Cosmos* (New York: Doubleday, 2004).

*Kalligas, D., Paul S. Wesson, and C. W. F. Everitt, "The Classical Tests in Kaluza-Klein Gravity," *Astrophysical Journal*, vol. 439, p. 548 (1995).

*Kaluza, Theodor, "Zur Unitätsproblem der Physik [On the Unity Problem in Physics]," *Sitzungsberichte der Preussischen Akademie der Wissenschaften*, vol. 54, p. 966 (1921).

Kirshner, Robert P., *The Extravagant Universe: Exploding Stars, Dark Energy and the Accelerating Cosmos* (Princeton, N.J.: Princeton University Press, 2002).

*Klein, Oskar, "Quantentheorie und fünfdimensionale Relativitäts-theorie (Quantum Theory and Five-Dimensional Relativity Theory)," *Zeitschrift für Physik*, vol. 37, p. 895 (1926).

Krauss, Lawrence, *The Fifth Essence: The Search for the Dark Matter in the Universe* (New York: Basic Books, 1989).

*Liu, H., and Paul S. Wesson, "Universe Models with a Variable Cosmological 'Constant' and a 'Big Bounce'," *Astrophysical Journal*, vol. 562, p. 1 (2001).

Livio, Mario, *The Accelerating Universe: Infinite Expansion, the Cosmological Constant and the Beauty of the Cosmos* (Hoboken, N.J.: Wiley, 2000).

Mach, Ernst, *Space and Geometry*, translated by Thomas McCormack (Chicago: Open Court, 1897).

Magueijo, Joao, *Faster than the Speed of Light: The Story of a Scientific Speculation* (Cambridge, Mass.: Perseus, 2003).

*Overduin, J. M., "Solar-System Tests of the Equivalence Principle and Constraints on Higher-Dimensional Gravity," *Physical Review D*, vol. 62, p. 102001 (2000).

*Overduin, J. M., and Paul S. Wesson, *Dark Sky, Dark Matter*, (Bristol: Institute of Physics Press, 2003).

Parker, Barry, *Invisible Matter and the Fate of the Universe* (New York: Plenum, 1989).

Penrose, Roger, *The Road to Reality: A Complete Guide to the Laws of the Universe* (New York: Knopf, 2005).

Rees, Martin, *Before the Beginning: Our Universe and Others* (Reading, Mass.: Perseus Publishing, 1998).

Rees, Martin, *Just Six Numbers: The Deep Forces That Shape the Universe* (New York: Basic Books, 2001).

Sagan, Carl, *Cosmos* (New York: Random House, 1980).

Seife, Charles, *Alpha and Omega: The Search for the Beginning and End of the Universe* (New York: Viking, 2003).

Singh, Simon, *Big Bang: The Origins of the Universe* (London: Fourth Estate, 2005).

Smoot, George, and Keay Davidson, *Wrinkles in Time* (New York: William Morrow, 1993).

Tyson, Neil deGrasse, and Donald Goldsmith, *Origins: Fourteen Billion Years of Cosmic Evolution*, (New York: Norton, 2004).

Webb, Stephen, *If the Universe Is Teeming with Aliens . . . Where Is Everybody? Fifty Solutions to Fermi's Paradox and the Problem of Extraterrestrial Life* (New York: Copernicus, 2002).

Weinberg, Steven, *The First Three Minutes: A Modern View of the Origin of the Universe* (New York: Basic Books, 1993).

*Wesson, Paul S., *Cosmology and Geophysics* (New York: Oxford University Press, 1978).

*Wesson, Paul S., "The Equivalence Principle as a Symmetry," *General Relativity and Gravitation*, vol. 35, p. 307 (2003).

*Wesson, Paul S., "A New Dark-Matter Candidate: Kaluza-Klein Solitons," *Astrophysical Journal Letters*, vol. 420, p. 49 (1994).

*Wesson, Paul S., "The Shape of the Universe," *Astronomy and Geophysics*, vol. 43, no. 12, p. 13 (2002).

*Wesson, Paul S., *Space-Time-Matter: Modern Kaluza-Klein Theory* (Singapore: World Scientific, 1999).

Index